T0328119

EEG-BASED DIAGNOSIS OF ALZHEIMER DISEASE

EEG-BASED DIAGNOSIS OF ALZHEIMER DISEASE

A Review and Novel Approaches for Feature
Extraction and Classification Techniques

Nilesh Kulkarni
Vinayak Bairagi

Academic Press is an imprint of Elsevier
125 London Wall, London EC2Y 5AS, United Kingdom
525 B Street, Suite 1800, San Diego, CA 92101-4495, United States
50 Hampshire Street, 5th Floor, Cambridge, MA 02139, United States
The Boulevard, Langford Lane, Kidlington, Oxford OX5 1GB, United Kingdom

Notices
Knowledge and best practice in this field are constantly changing. As new research and experience broaden our
understanding, changes in research methods, professional practices, or medical treatment may become necessary.

Practitioners and researchers must always rely on their own experience and knowledge in evaluating and
using any information, methods, compounds, or experiments described herein. In using such information
or methods they should be mindful of their own safety and the safety of others, including parties for whom
they have a professional responsibility.

To the fullest extent of the law, neither the Publisher nor the authors, contributors, or editors, assume any
liability for any injury and/or damage to persons or property as a matter of products liability, negligence or
otherwise, or from any use or operation of any methods, products, instructions, or ideas contained in the
material herein.

Library of Congress Cataloging-in-Publication Data
A catalog record for this book is available from the Library of Congress

British Library Cataloguing-in-Publication Data
A catalogue record for this book is available from the British Library

ISBN: 978-0-12-815392-5

For information on all Academic Press publications visit our website at
https://www.elsevier.com/books-and-journals

Working together
to grow libraries in
developing countries

www.elsevier.com • www.bookaid.org

Publisher: Mara Conner
Acquisition Editor: Chris Katsaropoulos
Editorial Project Manager: Mariana Kuhl
Production Project Manager: Sruthi Satheesh
Designer: Christian Bilbow

Typeset by Thomson Digital

To our families & friends,
Lord Ganesha whose blessings were worth &
who helps us see what's important and what's not

CONTENTS

Acknowledgments

It is a privilege for us to have been associated with Dr. P. B. Mane, the source of inspiration, during our research work and writing of this book. It is with great pleasure that we express our deep sense of gratitude to him for his valuable guidance, constant encouragement, motivation, support, and patience throughout this research work. His continuous inspiration helped lot for our personal development and shaped our career as a passionate researcher.

We would also like to thank Dr. D. K. Shedge and Dr. S. B. Dhonde for their valuable suggestions and moral support while carrying out the research work. We are also thankful to the reviewers of this book and other staff of Elsevier Publishing Corporation for their support and motivation.

We would specially like to thank Dr. Nilima Bhalerao, Assistant Professor, Smt. Kashibai Navale Medical College and General Hospital, Pune for providing necessary database and valuable information related to EEG signals and validating our results.

We wish to express our deepest sense of gratitude to our beloved parents, friends, and all family members for their moral support and blessings, which enabled us to complete this task. Our heartful thanks go to our family members for their patience, understanding, and cooperation during these days.

Finally, we wish to acknowledge Mariana Kühl Leme, Editor and Sruthi Satheesh, Project manager as well as Anita Mercy Vethakkan for their unusually great help and efforts during the period of preparing the manuscript and producing the book. Finally, we would like to thank all those who have helped directly or indirectly during the writing of this book.

Nilesh Kulkarni
Vinayak Bairagi

CHAPTER ONE

Introduction

Human Brain contains 10^{10} neurons [1]. In general, the thing which makes it unique is not the high number of cells but the ability to interact between them. It is well-known that the human body is controlled by human brain. In general, its study using neuroimaging techniques has represented a great advanced for science.

Neurodegenerative diseases are the group of disorders that affect the brain. They are basically related with changes in the brain that leads to the loss of brain structure including the death of some neurons [1,2]. The most well-known disease of this group includes Parkinson's disease, Alzheimer disease (AD), and Huntington's disease.

AD is the most prevalent neurodegenerative disease. As per the Neurologists reports, there is no cure for this disease. But, there are treatments that may delay the symptoms if they are provided in the first stages of the disease. Therefore, an early diagnosis of AD is a key issue for patients suffering from this disorder. Early diagnosis is difficult but the symptoms of diseases are confused with normal ageing effects. Due to this, Electroencephalography (EEG) has been presented as a useful technique that will facilitate the early diagnosis of AD. EEG is one of the imaging methods to study the brain activity. The economic price of EEG and its simplicity in use in comparison with other method make it a suitable choice for hospitals and research centers [1,2]. EEG records the brain signals using electrodes attached to the scalp. EEG recordings of AD patients show some characteristic changes that can be used as biomarkers of the pathology.

1.1 ALZHEIMER DISEASE

AD is a neurodegenerative and most prevalent form of age-related dementia in modern society. It affects behavioral and cognitive deficits. AD is positioned to become the scourge of this century bringing with its enormous social and personal costs [3,4]. It was discovered by Alois Alzheimer in 1906 over more than 100 years ago but research in this symptoms, causes, risk factors, and treatment has gained momentum in last 40–45 years. Even relevant aspects of AD are revealed, changes causing on AD patients is to be discovered. AD generally causes the loss of neurons in brain. It also damages neurons. Damaged neurons no longer function normally and may die. Dead neurons cannot be replaced once lost. By the time, brain cells shrinks dramatically, affecting all its functions. AD affects the patients in different ways, changing the rate of progression for each subject [3]. The initial symptoms of AD include the worsening ability to remember new information. This occurs due to the malfunctioning of neurons. As the neurons in different parts of brain regions die and malfunction, individuals experience other difficulties. The following listed are the different symptoms of AD [5]:

- Loss of memory; which interferes in daily life.
- Difficulties in solving problems.
- Results difficult to complete familiar tasks at home, at work, or at leisure.
- Poor judgment.
- Difficulties in remembering new words either speaking or writing.
- Confusion with time or place.
- Changes in personality or mood, includes apathy and depression.
- Withdrawal from work or societal activities.

AD is listed as the sixth leading cause of death in United States. It is also fifth leading cause of death for people of 65 years and older [6]. There includes a variety of parameters which are linked with incidence of AD, including age, gender, genetic factors, head injury, and Down syndrome. Experts believe that AD is caused by multiple factors than single causes. The major risks factor includes [3]:

1. Age: An advanced age is the greatest risk factor for AD. Even though age, is the greatest risk, is not sufficient to cause the disease.
2. Family history: Individuals with a familiar suffering AD are more likely to later develop AD.
3. APOE έ4 gene: Research studies estimate that between 40% and 65% of people diagnosed with AD have one or two copies of the APOE έ4 gene.
4. Mild cognitive impairment (MCI): Patients suffering from MCI are more likely to develop AD and other dementia than people without MCI. However, not all patients suffering from MCI latter develop AD. Therefore this is a key stage for studding AD.
5. Cardiovascular disease risk factor: It is suggested that the health of the brain is related with the heath of the heart and blood vessels. A good blood pleasure ensures that the brain receives the oxygen and nutrient necessary for its normal functioning.
6. Social and cognitive engagement: Some studies suggest that remaining mentally and socially active may reduce the risk of AD and other dementia. The exact mechanism underlying this situation is unknown [4].
7. Education: People with fewer years of formal education are at higher risk for AD and other dementia, than those with more year or formal education.
8. Traumatic brain injury: Moderate and sever traumatic brain injuries increase the risk of developing AD [5].

1.2 CAUSES AND SYMPTOMS OF THE DISEASE

AD does not present the same evolution in all patients. Rather it depends in part on the age of the pathology declaration and on the health conditions of a person, related with the risks factors as discussed above. It is believed that the brain change contributes to the development of AD which includes accumulation of the β-amyloid (Aβ) plaques and neurofibrillary tangles composed of tau amyloid fibrils. This accumulation of Aβ protein is produced outside the neurons, interfering with the communication of neurons, which happens in synapses and contributes to death of cells. The accumulation of tau angles is produced inside the neuron. In this manner, the tau tangles blocks the transport of nutrients to the neurons, which is referred to as death of cells. Currently, there is no cure for AD; but administering certain medications in initial stage (dementia) may delay the onset of symptoms [1–3].

Brain changes due to AD may begin 20 years before symptoms appear. At the beginning stage of the disease, patients are able to function normally. Afterwards, the brain can no longer compensate the Neuronal damage which has occurred. At this stage, the AD patients start to show subtle decline in cognitive functions. In this stage, the death of neurons is so significant such that patients start to present cognitive decline [3].

In later stage, basic functions are also impaired. The evolution of the disease is illustrated in Fig. 1.1. First, the accumulation of Aβ plaques and tau tangles starts in the hippocampus, the part of the brain where memories are first found (Fig. 1.1A). More and more regions of the brain are affected by the accumulation of Aβ plaques and tau tangles which comprises the brain function and presents the different stage of disease. At the next stage, the region of brain where language is processed is affected, compromising the ability of the patient to speak. Later, the disease affects the frontal part of the brain region where reasoning and planning is performed (Fig. 1.1B). The accumulation of plaques and tangles affects the part of the brain where emotions are regulated. At this stage, patients lose their control over moods and feelings. In advanced AD, cortex is seriously damaged (Fig. 1.1C). Individuals in this case lose their ability to communicate, to recognize family and to care for themselves.

1.3 STAGES AND CLINICAL DIAGNOSIS OF ALZHEIMER'S DISEASE

The progression of AD is classified into four stages. The first stage is termed as "MCI," that is, MCI. It usually presents some memory impairment. It retains their abilities in other cognitive domains and functional activities. Some MCI patients (6–25%) develop AD. The next stage is characterized by growing cognitive deficit. The second and third stage is termed as mild AD and moderate AD, while last stage is termed as severe AD; it is complete dependent on caregivers. Mild and moderate AD are key stages, an early diagnosis of AD in these stages can be done and proves beneficial since medications works in this stage [1,2]. By postmortem analysis of brain with dementia, a definitive diagnosis of AD can be made.

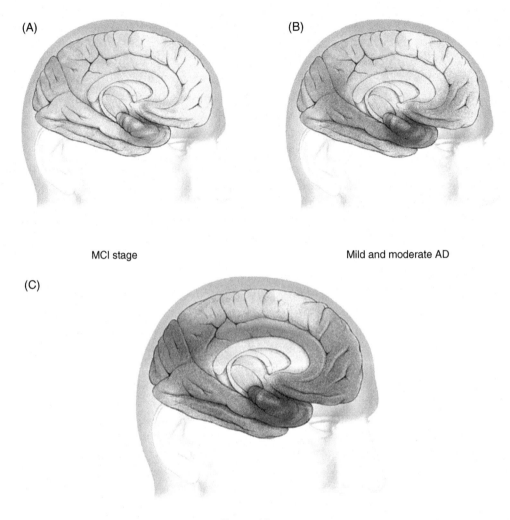

(A)

(B)

MCI stage

Mild and moderate AD

(C)

Severe AD stage

Figure 1.1 *Evolution of Alzheimer disease (AD) in the brain.* Plaques and tangles (shown in the dark gray shaded areas) tend to spread through the cortex in a predictable pattern as AD progresses. *(Courtesy: World Alzheimer Report 2014; M. Mattson, Pathways towards and away from Alzheimer's disease, Nature, 430 (2014) 631–639 [3]).*

A clinical diagnosis of AD is done on medical records, physical and neurological examination, laboratory tests, neuroimaging and Neurophysiological evaluation, such as Mini Mental State Examination (MMSE). Information from family members and close friends is also used as input. The diagnosis of AD is done by combining all the existing information; it is still difficult to diagnose the AD because the symptoms are often dismissed as normal consequence of ageing. Thus, AD is major health problem; it also causes tremendous suffering to the families of patients [1,2]. For the elderly, it is one of the most dreaded afflictions that threaten to rob them of their independence and dignity at the end of life [3,4].

1.4 IMPORTANCE OF DIAGNOSIS OF ALZHEIMER'S DISEASE AND ITS IMPACT ON SOCIETY

A positive diagnostic gives the patient and his family time to inform them about the disease, to make life and financial decisions related to the disease, and to plan for the future needs and care of the patients. A negative diagnostic may ease anxiety over memory loss associated with aging. It also allows for early treatments of reversible conditions with similar symptoms (such as thyroidal problems, depression, and nutrition or medication problems) [1–3]. Current symptoms-delaying medications have a given time frame during which they are effective. Early diagnosis of AD helps ensure prescription of these medications when they are most useful. Early diagnosis of AD also allows prompt treatment of psychiatric symptoms such as depression or psychosis, and as such reduces the personal and societal costs of the disease [3,4]. As research progresses, preventive therapies may be developed. Early diagnosis raises chance of treating the disease at a nascent stage, before the patient suffers permanent brain damage. Finally, as institutionalization accounts for a large part of health care costs incurred because of AD, by preserving patient's independence longer and preparing families for the needs of AD patients, timely diagnosis further decreases the societal cost of the disease [4].

As of 2013, there were an estimated 44.4 million people with dementia worldwide. This number will increase to an estimated 75.6 million in 2030, and 135.5 million in 2050. Much of the increase will be in developing countries. Already 62% of people with dementia live in developing countries, but by 2050 this will rise to 71%. The fastest growth in the elderly population is taking place in China, India, and their South Asian and western Pacific neighbors. In fact, the prevalence of AD is estimated to almost double in every 20 years, reaching a total of 65.7 million in 2030 and 115.4 million in 2050 [3]. This growth will be more significant in low and middle income countries [3] as shown in Fig. 1.2, and it is driven mainly by the demographic ageing and population growth. For instance, in regions such as Central Latin America, North America and Middle East, the expected growth between 2010 and 2050 is over 400% [3]. Also, the number of deaths related to Alzheimer's is still experiencing a marked increase, contrarily to other major causes whose numbers are declining [4,5]. This fact is not surprising since AD remains incurable and the number of people affected is not showing signs of slowing down. As a consequence, AD is already one of the most important causes of death, particularly in developed countries, ranking fifth in the United States for those aged 65 years or older. Fig. 1.3 compares changes in mortality between the years of 2010 and 2014 for several diseases.

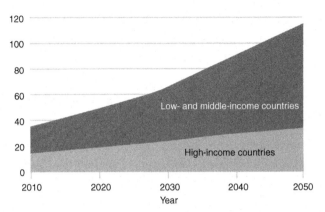

Figure 1.2 *Estimated number of patient with dementia (in millions) until 2050 in high income, low and middle income countries. (Courtesy: World Alzheimer Report 2014; M. Mattson, Pathways toward and away from Alzheimer's disease, Nature, 430 (2014) 631–639 [3]).*

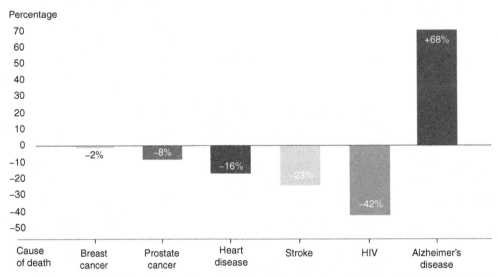

Figure 1.3 *Percentage changes in selected causes of death between 2010 and 2014 in the United States. (Courtesy: Alzheimer's Association Facts and Figures, 10 (2014) e47–e92 [5]).*

1.5 A BRIEF REVIEW ON DIFFERENT METHODS USED FOR DIAGNOSIS OF ALZHEIMER DISEASE

Studying the early diagnosis of AD using EEG signals encompasses different fields of knowledge. Therefore, this section explores previous works in these fields dividing them into three main parts. First, survey based on diagnosis of AD using neuroimaging is described in Section 1.5.1. Then in Section 1.5.2, diagnosis of AD using EEG signals is described, presenting the existing knowledge of this disease. Finally, this section examines the most relevant studies which specifically focus on early diagnosis of AD using EEG, presenting the main changes that AD causes on EEG signals and how this can be parameterized with different measures. In next part, that is, Section 1.5.3, we compare the different features used in the literature along with obtained classification accuracy. Thus, the literature has been reviewed and classified in the following categories:

- Diagnosis of AD by use of different Neuroimaging techniques
- Diagnosis of AD by use of EEG technique.

1.5.1 Role of neuroimaging based techniques in diagnosis of Alzheimer disease

Numerous clinical methods are extensively used for the diagnosis of AD, such as neuroimaging techniques, physiological markers, and genetic analysis. Neuroimaging is one of the well-accepted methods for definitive diagnosis of dementia. As we know that there is no cure for AD in final stage but early diagnosis of AD helps to save the life of the patient in early stage by giving proper medications. For diagnosis of neurodegenerative disease like dementia, information extracted from 3D brain images is the principal behind the development of automatic tool [6]. A Neuroimaging technique increases the confidence of diagnosis. Along with this, computer-aided diagnosis (CAD) system uses the discriminative value of brain images produced by different neuroimaging techniques, such as MRI, PET, and SPECT to identify the patient suffering from AD, MCI, Dementia, or normal.

Imaging has a key role in medical diagnosis, education and noninvasive therapeutics. The new scientific and technological advances boost the complex issues of diseases, such as Alzheimer, Epilepsy, and many more. The interaction of computers and technology made imaging a necessity due to possibility in diagnosis of disease along with providing the treatment. Medical imaging is useful whenever we need to diagnose the early condition of disease, planning of a surgery, or for radiation therapy. It involves the preprocessing of images, segmentation of images, features extraction and classification. Various aspects of medical image segmentations are discussed in [7]. Similarly, different features can be used for human–computer interaction for identifying the mental illness. Along with it different classifiers can be used, such as Support vector machines (SVM), neural networks (NN), and many more [8]. Medical imaging is regarded as a pillar in clinical, as well as medical applications. Computer tomography (CT) or magnetic resonance imaging (MRI), called conventional structural neuroimaging, plays a supportive role in diagnosis of memory disorder. Along with CT and MRI, emission tomography techniques also play a major role in clinical diagnosis of diseases. The emission based images termed as functional images represents the physiological functions with anatomical structure images providing information about physiological phenomenon. As previously stated CAD is a general tool used for a variety of application, such as to diagnose the disease in medical applications. CAD helps the physicians, researchers in diagnosing the disease in less time by identifying he patterns, making fewer efforts [9]. Let us review the contribution of different studies obtained in literature in diagnosis of AD by use of neuroimaging techniques.

Padilla et al. [9] presented a novel CAD technique for the early diagnosis of AD based on non-negative factorization and SVM. The proposed method is based on combination of non-negative matrix factorization for feature selection and reduction. They concluded that feature reduction steps provide a reduced set of variables representing the original data. The results of the proposed NMF-SVM method yielded 91% classification accuracy with high accuracy and specificity values. The limitation of this proposed methodology was the computational efficiency obtained was much lower and highly difficult to process the SPECT images. The solution for the above limitation is the future challenges in their research work.

Zhang et al. [10] proposed a technique for diagnosis of AD using structural MRI and hybrid classification system for distinguishing NC, MCI, and AD patients. They used PCA to reduce the dimensionality of feature vector of MRI data and resultant principal components retained important information. The kSVM-DT method was used as a classifier which gathers the principal components from MRI data and additional data and gave 80% classification accuracy. The limitation of their research work was classifier establishes machine oriented rules and not the human oriented rules. Other limitation is classification accuracy that is much poor for AD diagnosis. The future scope of their research work is to increase the classification accuracy using different classifiers, such as ANN, Bayesian Classifier, and Hidden Markov Models. The future work also includes to extract the more efficient features of MRI images and to include the phase image of structural MRI data.

Freebrough et al. [11] used MR texture as a diagnostic tool, marker, and a measure of progression in AD. They demonstrated that a texture discriminant function derived from MRI brain scans yields significantly different values for AD patients compared to normal controls. They concluded that such measures can be useful in the diagnosis of AD. Their study ignored the effects of texture changes associated with motion artifact. The texture discriminant function developed in the research work is based upon two-dimensional texture measures calculated over 3D structure. Their research findings concluded that results obtained in their research work were not much concise and there is a need to obtained more precised results that is much different to obtain.

Liillemark et al. [12] investigated a marker based on procrustes aligned center of masses and the percentile surface connectivity between regions. The markers used in the above were classified using linear discriminant analysis (LDA) in a cross validation setting and comparing with whole brain and hippocampus volume. They concluded that relative proximity marker based on the structural MRI includes more information. They also concluded that proximity markers have the potential to assist in early diagnosis of AD they also concluded that structural MRI is a suitable modality for detection of AD and AD progression. Along with this, they also suggested that additional information can contribute to the refinement of the AD markers and can be able to give more detail picture of AD diagnosis and progression.

Rueda et al. [13] presented an automatic image analysis method that reveals discriminative brain patterns, which are associated with presence of neurodegenerative diseases which grades out any neurological disease. They suggested that this is accomplished by a fusion strategy which mixes together bottom–up and bottom–down information flow. They have used OASIS-MIRIAD database for evaluating their results. They also suggested that bottom–up information consists of multi scale analysis of different image features and top-down stage is the learning and fusion strategies formulated as a max margin multiple kernel optimization problems. They proposed that with adequate and exhaustive evaluation in large datasets consisting of sufficient amount of AD stages, above method can be used for second diagnostic opinion in the clinical use. The disadvantage of their work was the feature computation in proposed algorithms was mathematical and requires more time for computations.

Illan et al. [14] proposed a spatial component (SC) based approach for diagnosis of AD focusing on MRI images. They subdivided the MRI images in different spatial components and modeled the dependencies between affected regions using Bayesian network. They obtained the 80% accuracy in their present work with the conclusion that modeling the dependences between the components increased the recognition of different patterns of brain degeneration in AD.

Wan et al. [15] proposed a sparse Bayesian learning algorithm for MRI images based on a sparse multivariate regression model. The proposed algorithm offered a flexibility to examine the non-linear relationship between responses and predicted database. They illustrated the superiority of the proposed algorithm by applying it to the prediction of cognitive scores of subjects from their MRI measures. The proposed algorithm showed the highest prediction accuracy and also demonstrated the ability to accurately identify imaging biomarkers. But, the time taken to compute the algorithm is high.

Morgado [16], "Automated Diagnosis of Alzheimers Disease using PET images: a study of alternative procedures for feature extraction and selection," focused on PET images and studied the alternatives methods for computerized diagnosis systems of AD using the features extraction and selection techniques. He studied FDG-PET image for different scales, resolution sand texture descriptor. He also proposed local binary pattern (LBP), a novel method for 3D image diagnosis. Other conventional methods based on correlation measures and mutual information was also studied. Classification was done using SVM classifier to distinguish between AD patients, MCI patients and control group based on dichotomous fashion.

Ledig et al. [17] presented a classification frame based on the different set features extracted from structural MRI images which are examined with respect to their discriminative abilities. The above features were based on volumetric and morphologic parameters, intensity of images and patch similarities. They used a three class classification for classification between AD versus MCI versus normal on which they obtained 59% accuracy on the above feature sets. The increase in classification accuracy was the objective of their future work.

Liu et al. [18] outlined that the bottleneck exists in the diagnostic performance method due to the lack of efficient strategies. They proposed a novel framework for diagnosis of AD with deep learning architecture. This framework distinguishes the four stages of AD. They have incorporated the unsupervised feature representation in their work flow. Due to the use of stacked auto-encoders, they obtained high-level features. The present system also uses zero masking strategy for data fusion to extract complimentary information from multiple data modalities. The proposed method was capable of fusing multi-modal neuroimaging features and has potential to require less labeled data. The authors assumed that synergy between different biomarkers can be further extracted with more training samples. They observed that the methodology had high accuracy on training data but low on testing database. The increase in same was objective of their future work.

From the above survey, different neuroimaging techniques such as MRI, PET, SPET plays a significant role in diagnosis of neurological disorder like Dementia and Alzheimer. But, it poses certain disadvantages. From above survey, it is to conclude that diagnosis accuracy of MRI is less and cannot extract efficient features (information) easily. Although different methods, such as MRI proximity markers exist; they are highly complex and costly. On the other hand, SPECT images gives a high accuracy but their computational efficiency is much poor and they are difficult to process. EEG can provide efficient information for diagnosis of AD. We can easily observe the abnormalities seen in the EEG signals of AD patients. Different abnormalities, such as slowing of EEG signals, perturbations in synchrony measures and reduced complexity of EEG signals are some of the abnormalities seen. Different EEG signals features play a significant role in AD diagnosis and can be used as a tool for classification. Different features of EEG signals, such as spectral, amplitude, coherence, phase-based features provides efficient information. These features can be easily classified through the use of different classifiers, such as SVM, LDA, and NN, etc. Due to the above shortcomings in the neuroimaging based techniques we focus on EEG based biomarkers for diagnosis of AD in early stage. As EEG is cheap and it reveals out the necessary information from graphical representation and the behavior of the signals. We concentrate on EEG signals for developing an automated system for diagnosis by means of employing various signal processing tools and machine learning methods. The brief survey based on EEG diagnosis of Alzheimer's is discussed later.

1.5.2 Role of electroencephalogram techniques in diagnosis of Alzheimer disease

As seen in previous section, neuroimaging technique plays an important role in diagnosis of neurodegenerative diseases, such as dementia, Alzheimer. Neuroimaging techniques are well-accepted methods for diagnosis of AD. But, it has certain disadvantages. The major problem of neuroimaging techniques includes the radiation risks. Apart from this, neuroimaging techniques, such as MRI, SPECT, PET, fMRI, and others are found to be much expensive, time consuming, and inconvenient. In such cases, non-neuroimaging techniques play a vital role in diagnosis of diseases. The advantages of non-neuroimaging techniques includes that they are inexpensive, repeatable, and can be easily used at home through wireless networks. In future non-neuroimaging methods, such as EEG can also be made portable [19].

Non-neuroimaging techniques, such as biomarkers, EEG are widely accepted methods for diagnosis of AD. Cerebrospinal fluid (CSF), plasma biomarkers, and genetic biomarkers are some of the biomarkers used in diagnosis of dementia and AD [6]. The core diagnostic CSF markers for diagnosis of AD and dementia include Aβ42, T-tau, and P-tau. According to Blennow et al. [20], the low-level concentration of Aβ42 and high level of P or T − tau identify AD with sensitivity and specificity of 80%. Other biomarkers, such as genetic biomarkers also play a major in diagnosis of AD. Amyloid precursor protein (APP), Presenilin-1(PSEN1), and Presenilin-2 (PSEN2) are well-accepted susceptibility genes for early diagnosis of AD. Above biomarker and imaging techniques have limitations to their use although they have much development. The reason is due to the collection of CSF is difficult and uncommon for diagnosis of dementia and AD. It is also an invasive method. EEG is one of the tools which can be used for early diagnosis of AD. EEG is noninvasive, repeatable and can be easily evaluated. It is a personalized medical tool. EEG can be used for direct correlation of brain function, which can be used for monitoring brain activity. Non-linear analysis of EEG data shows the unique features to reveal the diagnosis of neurological diseases, such as Alzheimer, Epilepsy, and Parkinson's [21,22].

As compared to the other neuro-imaging methods, EEG has high-temporal resolution and as a consequence several abnormalities are found in the EEG of the patients. EEG is a non-invasive tool for diagnosis of neurological disease. The advantages of EEG are discussed in above section. Certain abnormalities are found in EEG signals of patients suffering from AD [23–27]. The various abnormalities found in patients suffering from AD include (1) slowing of EEG signals, (2) reduced complexity of EEG signals, and (3) perturbation in EEG synchrony. It is important to focus on these three abnormalities for early diagnosis of AD.

Let us review the work done by different researchers in field of AD diagnosis using EEG signals.

Jeong [28], "EEG dynamics in patients with Alzheimer's disease," states that AD is the most common neurodegenerative disorder characterized by cognitive and intellectual deficits and behavior disturbance. EEG has been used as a standard tool for diagnosing the AD in decades. Further, author has observed that, hallmark of EEG Abnormalities in AD patients are a shift of the power spectrum to lower frequencies and a decrease in coherence of fast rhythms. The paper reviews the important search of the EEG abnormalities in AD patients obtained from conventional spectral analysis and nonlinear dynamical methods. The abnormalities in the AD patients through EEG are characterized by the slowed mean frequency, less complex activity, and reduced coherences among cortical regions. The future work as proposed by this author is associated with the improvement of the accuracy of differential diagnosis and early detection of AD based on multimodal approaches, longitudinal studies on nonlinear dynamics of the EEG, drug effects on the EEG dynamics, and linear and nonlinear functional connectivity among cortical regions. He has concluded in his review that the diagnostic accuracy of the EEG based on a broad survey of the literature is currently about 80%.

Rodriguez et al. [29], "Spectral Analysis of EEG in Familial Alzheimer's Disease with E280A Presenilin-1 Mutation Gene," states that the goal of the presented study was to determine the possible impact of spectral EEG analysis to detect early functional changes in preclinical stages of familial AD. They had selected the three subjects for diagnosis with hereditary AD such as a Probable AD group, an Asymptomatic Carrier (ACr) group and a normal group. Their results show that there is presence of beta-bands alteration in ACr groups in the absence of clinical sign. Fast frequency bands change mainly in gamma-frequencies, in people with clinical sign of MCI. They also suggested that a reduction of beta power is not only due to ageing, but may reflect an alteration of AD especially in the early stage.

Dauwels et al. [30], "Diagnosis of Alzheimer's disease from EEG Signals: Where Are We Standing?." have shown that MCI and AD cause EEG signals to slow down and MCI and AD are associated with increase of power in slow frequencies (delta- and theta-band) and a decrease of power in fast frequencies (alpha- and beta-band). Nevertheless, increased gamma-band power has been reported in MCI and AD patients compared to healthy age–matched control subjects. The authors also concluded that slowing of EEG signals, perturbations in synchrony measures and reduced complexity of EEG signals are the abnormalities seen in patients with AD. The authors conclude that EEG plays a key role in diagnosis of AD.

Bock et al. [31], "Early Detection of Alzheimer's Disease Using Nonlinear Analysis of EEG via Tsallis Entropy," had worked on the Early Detection of AD using non-linear analysis of EEG via Tsallis entropy. Their study focuses on participants with MCI and includes both a delayed visual match-to-sample (working memory) task and a counting backwards task (eyes closed) for comparison. The EEG data were quantified using Tsallis entropy, and the brain regions analyzed included the prefrontal cortex, occipital lobe, and the posterior parietal cortex. They concluded that Tsallis entropy-based EEG analysis has been shown to be a highly promising potential diagnostic tool for MCI and early dementia.

Fraga [32], "Towards an EEG-Based Biomarker for Alzheimer's Disease: Improving Amplitude Modulation," had proposed an EEG-based biomarker for automated AD diagnosis, based on extending proposed "percentage modulation energy" (PME) metric. It was done basically to improve the signal-to-noise ratio of the EEG signal, PME features were averaged over different durations prior to classification. Two variants of the

PME features were Developed: the "percentage raw energy" (PRE) and the "percentage envelope energy" (PEE). Results are achieved with a support vector classifier when discriminating between healthy and mild AD patients. Thus, this paper proposes the spectro-temporal EEG-based biomarker for AD detection.

Poil et al. [33], "Integrative EEG biomarkers predict progression to Alzheimer's disease at the MCI stage," had used the integrative EEG biomarkers to predict the diagnosis of AD at MCI stage AD is a devastating disorder of increasing prevalence in modern society. MCI is considered a transitional stage between normal aging and AD. They predicted that EEG biomarker would provide anon-invasive and relatively cheap screening tool to predict conversion to AD. Their results showed that exploratory data mining and integration of multiple biomarkers might yield many exciting results on the large databases of neuroscience data build. By integrating six EEG biomarkers into a diagnostic index using logistic regression (LR) the prediction, sensitivity of 88% was achieved. Improvement in algorithms used for pre-selecting biomarkers was a future scope for their research work.

Dauwels et al. [34], "Slowing and Loss of Complexity in Alzheimer's EEG: Two Sides of the Same Coin?," had proposed that EEG of Alzheimer patients is slower. They have concluded that there exists the strong correlation between slowing and loss of complexity in two independent EEG datasets; which are (a) EEG of pre dementia patients MCI and control subjects; (b) EEG of mild AD patients and control subjects. They also investigated that the potential of EEG slowing and loss of EEG complexity as indicators of AD onset. Relative power and complexity were used as the features to classify the MCI and Mild AD patients versus age-matched control subjects. After the combination of the above two measures, classification rates of 83% (MCI) was obtained.

Gallego-Jutgla et al. [35], "Diagnosis of Alzheimer's Disease from EEG by Means of Synchrony Measures in Optimized Frequency Bands," had proposed the method to detect the AD by means of synchrony measures in optimized frequency bands. They presented the frequency band analysis of AD EEG signals with the aim of improving the diagnosis of AD using EEG signals. Several synchrony measures were assessed through statistical tests (Mann–Whitney U test); including correlation, phase synchrony and Granger causality measures. They investigated that the frequency range (5–6 Hz) yields the best accuracy for diagnosing AD, which lies within the classical theta band (4–8 Hz). Their results predicted that EEG of AD patients is more synchronous than in healthy subjects within the optimized range 5–6Hz, which is in sharp contrast with the loss of synchrony in AD EEG reported in many earlier studies. Furthermore, they are working on the large datasets to verify the effectiveness of the proposed approach. The ultimate objective of their research is to determine the most appropriate EEG frequency bands for diagnosing AD (and potentially other neurodegenerative diseases) with synchrony measures.

Cassani et al. [36], "The effect of automated artifact removal algorithms on EEG based AD diagnosis" had proposed the algorithm for automatic artifact removal (AAR) techniques for electroencephalograph based diagnosis of AD. They have evaluated the different AAR algorithms for diagnosis performance. Their results showed that wavelet enhanced ICA AAR algorithms works better for the investigated features, such as Spectral, amplitude, coherence, phase, and spectral modulation–based EEG features. The future scope of their research work is to focus more on exploration of AAR algorithms with studies involving EEG with 64 channels, EOG and fully automated statistical thresholding for EEG artifact rejection.

Trambaiolli et al. [37], "EEG Spectro-temporal Modulation Energy: A new feature for automated Diagnosis of AD," proposed a new feature for automated diagnosis of AD. They proposed a non-invasive AD diagnosis tool which characterizes EEG sub-band modulation. The proposed feature spectro-temporal modulation energy measures the rate with which each sub band is modulated. Modulated energy features were computed for different EEG signals including bipolar signals. SVM, LR, and NN were the classifiers used for classification purpose. The focus of ongoing work is to focus more on automated artifact removal by means of independent component analysis (ICA). Their future work is also to compare different linear and non-linear techniques for EEG based diagnosis of AD.

Staudinger et al. [38], "Analysis of Complexity Based EEG features for diagnosis of AD," studied and analyzed EEG signals using several non-linear signal complexity measures for AD diagnosis. The various measures include spectral entropy), spectral centroid (SC), spectral roll-off (SR), and zero-crossing rate, etc. They also used a *t*-test to determine the features that provides significant difference between the different groups. Their future works includes optimizing the classifiers as well as the other feature sets such as frame and epoch lengths, overlap between frames and combining them with wavelet based features.

Tsolaki et al. [39], "Electroencephalogram and Alzheimer's Disease: Clinical and research Approaches," suggested that AD is a neurodegenerative disorder which is characterized by cognitive deficits, problems in activities of daily living, and behavioral disturbances. Electroencephalogram (EEG) has been demonstrated as a reliable tool in dementia research and diagnosis. They also suggested that application of EEG in AD has a wide range of interest because EEG contributes to the differential diagnosis and the prognosis of the disease progression. Additionally such recordings of EEG data can add important information related to the drug effectiveness. The team tried to focus on the main research fields of AD via EEG and recent published studies. They also suggested that EEG is proved to be a reliable diagnostic tool in dementia research. They also discussed that AD is a cortical dementia in which EEG rhythms abnormalities are more frequently shown, whereas subcortical dementia exhibits relatively normal EEG patterns. They proved that EEG with contemporary statistical methods seems to be a reliable method to classify the clinical cases of cognitive impairment.

Akrofi et al. [40], "Classification of Alzheimer's Disease and Mild Cognitive Impairment by pattern recognition of EEG Power and Coherence," proposed a methodology to classify AD and MCI with high accuracy using EEG data. They used sequential forward floating search to select features from relative average power for channel locations in frequency bands delta, theta, alpha, and beta, and coherence between intrahemispheric channel pairs for the same frequency ranges. They obtained the 90% classifier accuracy when classifying MCI patients and normal subjects for selected feature sets. The obtained results showed that selecting features from a combined set of power and coherence features produced better results than the use of either feature independently. The combined feature set also showed better classification accuracy than a Bayesian classifier fusion approach in their study. The future scope in their work includes the use of a larger number of subjects and a theoretical determination of the probability distributions for which Bayesian data fusion would lead to an increase in OCRs.

Dauwels et al. [41], "On the Early Diagnosis of Alzheimer's disease from EEG Signals: A Mini Review," pointed out several challenges and topics for future research in Alzheimer diagnosis. They also suggested that it is hard to systematically benchmark and assess the existing methods for diagnosing AD from EEG signals. They also gave importance to investigate whether EEG helps to distinguish between MCI and different stages of AD and between AD and other dementias. They also predicted that EEG signals may be more or less discriminative for MCI and AD; a systematic exploration of different recordings conditions with the aim of diagnosing MCI and AD needs to be conducted. Their study also focuses to investigate the effect of medication and therapy on the EEG of AD patients.

Polikar et al. [42], "An Ensemble based Data fusion Approach for Early Diagnosis of Alzheimer's Disease," described an ensemble of classifiers based data fusion approach to combine information from two or more sources, which opts to contain complementary information, for early diagnosis of AD. Their emphasis was on sequentially generating an ensemble of classifiers that explicitly seek the most discriminating information from each data source. The application presented in work seeks the diagnostic identification of AD versus normal patients based on their ERP recordings. They used the event related potentials (ERPs) recorded from the Pz, Cz, and Fz electrodes of the EEG, which were decomposed into different frequency bands using multiresolution wavelet analysis. Their proposed data fusion approach includes generating multiple classifiers trained with strategically selected subsets of the training data from each source, which were then combined through a modified weighted majority voting procedure. They also presented implementation details and the promising outcomes of their current work. Their future work includes repeating virtually all experiments for more patients. They will be expanding this analysis to include a third cohort: patients suffering from Parkinson's disease, about 30% of whom eventually develop dementia. Their future scope also includes formal analysis of the algorithm on several different scenarios of data fusion and additional EEG channels.

Woon et al. [43], "Techniques for Early Detection of Alzheimer's disease Using Spontaneous EEG recordings," they proposed a set of novel techniques which helped to perform the task, and presented the results of experiments conducted to evaluate the approaches. The challenge in their work was to discriminate between spontaneous EEG recordings from two groups of subjects: one afflicted with MCI and eventual AD and the other an age-matched control group. The classification results obtained indicated that the proposed methods are promising additions to the existing tools for detection of AD. As such, it is possible that the features tested are markers of MCI and not of the future AD. Their future work involves the application of proposed algorithms for more number of databases involving the different types of subjects used.

Ahmadlou et al. [44], "Wavelet Synchronization Methodology: A New Approach for EEG Based Diagnosis of ADHD," presented multi-paradigm methodology for EEG based diagnosis of attention-deficit/hyperactivity disorder (ADHD) through adroit integration of nonlinear scientific techniques involving; wavelets, a signal processing technique; and NNs, a pattern recognition technique. The selected nonlinear features are generalized synchronizations known as synchronization likelihoods (SL). The methodology consists of three parts: first detecting the more synchronized loci (group 1) and loci with more discriminative deficit connections (group 2). In next part, SL's were computed, not of all electrodes, but between loci of group 1 and loci of group 2 in all sub-bands and the band-limited EEG. This part leads to more accurate detection of deficit connections. In concluding section, a classification method, radial basis function NN, is used to separate ADHD from normal subjects. The radial basis function NN classifier yielded a high accuracy of 95.6% for diagnosis of the ADHD in the feature space. To the best of the authors' knowledge this is the first time chaos theory and nonlinear science has been used for diagnosis of ADHD. Another novelty in their proposed methodology includes nonlinear features are detected not only from the full-band EEG, but also from EEG sub-bands.

The objective of Cichocki et al. [45], "EEG filtering based on blind source separation (BSS) for Early Detection of Alzheimer's Disease," was to develop a preprocessing method for EEG signal for improvement of detection of AD. The technique used in the present study includes filtering of EEG data using blind source separation and projection of components which are possibly sensitive to cortical neuronal impairment found in early stages of AD. They used EEG Relative power based features to classify between two groups. They also incorporated the use of LDA classifier for classifying the EEG data between two groups. The proposed algorithm increased the diagnosis as well as classification accuracy using the above method. Sensitivity and specificity were also calculated and they also showed the improved results. They concluded filtering based on BSS technique can improve the performance of the existing EEG approaches for early diagnosis of AD. It can also have potential for improvement of EEG classification in other clinical areas or fundamental research. The developed method is very general and flexible and can be allowed for various extensions and improvements.

Dauwels et al. [46], "A Comparative Study of Synchrony Measures for Early Diagnosis of Alzheimer's Disease Based on EEG," studied various synchrony measures in the context of AD diagnosis, including the correlation coefficient, mean-square and phase coherence, Granger causality, phase synchrony indices, information-theoretic divergence measures, state space based measures, and the recently proposed stochastic event synchrony measures. Their experimental results showed that EEG data show that many of those measures are strongly correlated (or anti-correlated) with the correlation coefficient, and hence, provide little complementary information about EEG synchrony. Measures that are only weakly correlated with the correlation coefficient include the phase synchrony indices, Granger causality measures, and stochastic-event synchrony measures. Their classifier yielded out classification rate of 83%. They also suggested that classification performance may be further improved by adding complementary features from EEG; this approach may eventually lead to a reliable EEG-based diagnostic tool for MCI and AD.

Vialatte et al. [47], "Early Detection of Alzheimer's Disease by Blind Source Separation, Time Frequency Representation and Bump Modeling," reported a novel application of BSS combined with time frequency representation and sparse bump modeling for the automatic classification of EEG data for early detection of AD and dementia. The developed system was applied to EEG recordings which were analyzed previously with standard feature extraction and classification methods. Comparing with the previous analysis, improved results were achieved, such as the overall correct classification rate were increased from 80% to 93% (sensitivity 86:4% and specificity 97:4%). The present study provided exciting prospects for early mass detection of the disease. They also suggested that the EEG-based diagnosis method is very cheap as compared to PET, SPECT, and fMRI scans, requiring only a 21 channel EEG apparatus. Sparse bump modeling proved to be a useful tool for compressing information contained in EEG time–frequency maps for diagnosis and classification. They research findings also concluded that bump modeling can be a good approximation of bursts and sufficiently well follow for important features of amplitude modulation of EEG oscillations and therefore it can become a promising way of compressing information contained in EEG and can be widely used for its analysis. The future scope of their work is to observe the dynamics of the bumps and the brain functional connectivity on large number EEG datasets with more number of subjects.

Lehmann et al. [48], "Application and Comparison of Classification Algorithms for Recognition of Alzheimer's disease in Electrical Brain Activity," explored that effective appliance of treatment strategies is essential for early detection of subjects with probable AD. In the present study, they explored the ability of a multitude of linear and non-linear classification algorithms to discriminate between the electroencephalograms (EEGs) of patients with varying degree of AD and their age-matched control subjects. Absolute and relative spectral power, distribution of spectral power, and measures of spatial synchronization were the different features used in their present work. Different classification algorithms were applied such as principal component linear discriminant analysis (PC LDA), partial least squares LDA (PLS LDA), principal component logistic regression

(PC LR), partial least squares logistic regression (PLS LR), bagging, random forest, SVM, and feed-forward NN. In the present study it was seen that SVM and NNs showed a slight superiority rather more classical classification algorithms performed nearly equally well. Using random forests classification technique a sensitivity of 85% and a specificity of 78% was obtained, whereas for the comparison of moderate AD versus controls, using SVM and neural NN, sensitivity of 89% and specificity of 88% were achieved.

Jacques et al. [49], "Multiresolution Wavelet Analysis and Ensemble of Classifiers for Early Diagnosis of Alzheimer Disease," suggested that diagnosis of AD at an early stage is essential due to growing number of the elderly population affected and the lack of a standard and effective diagnosis tools available to community healthcare providers. They emphasized that recent studies have used wavelets and other signal processing tools to analyze EEG signals in an attempt to find a non-invasive biomarker for AD and dementia and most studies showed a varying degrees of success. They also suggested that the previous studies have used automated classifiers such as NN; but however the use of an ensemble of classifiers has not been previously explored in the research findings but may prove to be beneficial in future. In the present study, multi-resolution wavelet analysis was performed on ERPs of the EEG which were then used with the ensemble of classifiers based on Learn + algorithm. They obtained 88% classification accuracy in their work.

Vialatte et al. [35], "Blind Source Separation and Sparse Bump Modeling of Time Frequency representation of EEG Signals: New tools for Early Detection of Alzheimer Disease," proposed a novel method for early detection of AD using electroencephalo-graphic (EEG) recordings. Firstly, BSS algorithm was applied to extract the most signifi-cant spatio-temporal components; these components were then subsequently wavelet transformed; the resulting time frequency representation was approximated by sparse "bump modeling"; whereas in the final step reliable and Discriminant features were then selected by orthogonal forward regression and the random probe method. These features were fed to a simple NN classifier. The present method was applied to EEG recorded in patients with MCI who later developed AD, and in age-matched controls. This pres-ent study lead to a substantially improved performance (93% correctly classified, with improved sensitivity and specificity) over classification results previously published on the same set of data. The method is expected to be applicable to a wide variety of EEG classification problems.

Gallego-Jutglà et al. [50], "Diagnosis of Alzheimer Disease from EEG by means of Synchrony Measures in Optimized frequency Bands," presented a frequency band analy-sis of AD EEG signals with the aim of improving the diagnosis of AD using EEG signals. In the present work, multiple synchrony measures were assessed through statistical tests (Mann–Whitney U test), including correlation, phase synchrony, and Granger causality measures. LDA was conducted with those synchrony measures as features for classifica-tion. They highlighted that the frequency range (5–6 Hz) yields the best accuracy for diagnosing AD, which lies within the classical theta band (4–8 Hz). The corresponding classification error obtained was 4.88% for directed transfer function Granger causality measure. Their results obtained were also interesting which showed that EEG of AD patients is more synchronous than in healthy subjects within the optimized range 5–6 Hz, which is in sharp contrast with the loss of synchrony in AD EEG as reported in many earlier studies. This new finding from their work provided new insights about the neurophysiology of AD. Additional testing on larger AD datasets is required to verify the effectiveness of the proposed approach of their present work.

Ghorbanian et al. [51], "Identification of Resting and Awake state EEG features of Alzheimers Disease Using Discrete Wavelet Transform," proposed that AD is associated with deficits in cognitive processes and executive functions. They also suggested that power spectrum-based features have been characterized with montage recordings and conventional spectral analysis during resting eyes closed and eyes open conditions. In the present study, EEG signal was recorded from left prefrontal cortex lobe. The signal were then decomposed into sub bands corresponding to the major brain frequency bands using discrete wavelet transform and different statistical features were calculated from each band. For the purpose of classification, decision tree algorithm was applied along with multivariate and univariate statistical analysis to identify predictive features across resting and active states. They concluded that EEG data recorded during resting conditions is able to differentiate between AD and control subjects.

Kang et al. [52], "Principal Dynamic Mode Analysis of EEG Data for assisting the Diagnosis of AD," examines the model of the causal dynamic relationships between frontal and occipital EEG time-series recording of AD patients versus control subjects to reveal reliable differentiating characteristics. Their proposed modeling approach utilizes the concepts of PDM's and the associated non-linear functions (ANF's). They used different features, such as elative EEG power (RP), median frequency, sample entropy, spectral coherence to distinguish the EEG subjects of both groups. The results concluded that ANF's of two PDM's; corresponding to the delta-theta and alpha bands can delineate the two groups.

Traimboli et al. [53], "Does EEG Montage Influence AD Electro clinic Diagnosis," suggested that there is no specific AD diagnostic test. They concluded that AD diagnosis relies on clinical history, neurophysiologic and laboratory tests, including EEG and neuroimaging tests. In this sense, new approaches and techniques are necessary for accurate diagnosis and to measure the diagnosis results. They also suggested that quantitative EEG (qEEG) can be used as a diagnostic tool for AD diagnosis. The aim of their study was to answer if any distinct electrode montages have sensitivity when differentiating controls from AD patients. They analyzed EEG spectral peaks of different bands, such as Delta, Theta, Alpha, Beta and Gamma) and also compared reference electrodes, such as Biauricular, Longitudinal Bipolar, cross bipolar, Counterpart Bipolar, and Cz reference. SVM and logistic regression was the classifier used which showed that Counterpart Bipolar Montage is most sensitive electrode combination (Table 1.1).

From the above literature survey, it is observed that different researchers worked on EEG signals for diagnosis of AD and Dementia. Different signal processing techniques were applied to signals for diagnosis of AD from which features were computed for

Table 1.1 Survey on accuracy obtained by different researchers worldwide in EEG-based Alzheimer disease diagnosis

Sr. No.	Authors	Proposed method	Accuracy (%)
1.	Simon-Shlomo et al. [33]	Integrative EEG biomarkers to predict the diagnosis of Alzheimer disease at MCI stage	88
2.	Dauwels et al. [30]	They have used the relative power and complexity measures as features to classify the MCI and mild AD. After the combination of two synchrony measures (Granger causality and stochastic event synchrony) classification rates were obtained	83
3.	Jeong et al. [28]	The paper reviews the important search of the EEG abnormalities in AD patients. They have used the conventional spectral analysis and nonlinear dynamical methods to find the abnormalities in EEG signal	80
4.	Bock et al. [31]	Non-linear analysis of EEG via Tsallis entropy	82
5.	Cassani et al. [36]	Used spectral coherence, phase and spectral modulation–based features for diagnosis of the three groups: Alzheimer patients, MCI group, and normal group	89
6.	Staudinger et al. [38]	Non-linear features such as Higuchi's fractal dimension, spectral entropy, etc.	70
7.	Polikar et al. [42]	Multiresolution wavelet analysis for analysis of EEG signals	85
8.	Ahmadlou et al. [44]	Wavelet-based signal processing approach for analysis of EEG signals for distinguishing between Alzheimer's disease and normal subjects	90.6
9.	Lehmann et al. [48]	Absolute and relative power, measures of spatial synchronization	89
10.	Jacques et al. [49]	Multiresolution wavelet analysis for signal analysis for distinguishing between the two groups	88

Note: The classification accuracies presented are different for different databases, as EEG database of Alzheimer's disease is not available in public domain.

distinguishing two subjects. The features were then selected and classified by means of suitable classifier. From the Table 1.1, we can observe that many researchers focused on time–frequency–based features as the means for diagnosis. From the table, we can observe that, many researchers also focused on EEG relative power, coherence-based features, phase synchrony features in frequency domain. Complexity-based features, including entropy-based feature were also used. Sample entropy and spectral entropy were used as the features for diagnosis. Similarly, Granger causality, phase synchrony, and stochastic event synchrony were also considered as the complexity-based features for diagnosis. Wavelet based features were also studied for diagnosis of AD. Different wavelets, such as Daubechies and Morlet were studied for diagnosis by decomposing it at certain level to extract statistical features from it. In our study, the main aim is to study the above features for diagnosing the disease in early stage with increased diagnostic accuracy (up to 95%). We have incorporated the use of new complexity based features in our study for diagnosis of AD. We will study these new features in further chapters.

SUMMARY

This chapter presented an introduction to AD. Various causes and symptoms of AD are discussed. It is to highlight that there is no cure for AD in final stage but early diagnosis of AD helps to save the life of the patient in early stage by giving proper medications and delay the disease. Different neuroimaging, such as MRI, SPECT, etc., and non-neuroimaging, such as EEG are also discussed in this chapter for AD diagnosis. In further chapters, we will discuss the role of EEG in clinical use and its importance in AD diagnosis.

REFERENCES

[1] Nunez PL, Srinivasan R. Electric Fields of the Brain: The Neurophysics of EEG. 2 New York: Oxford University Press; 2006.
[2] Kropotov JD. Quantitative EEG Event-Related Potentials and Neurotherapy. Brazil: Elsevier Inc.; 2009.
[3] Mattson M. Pathways towards and away from Alzheimer's disease. Nature 2004;430:631–9.
[4] Meek PD, McKeithan K, Shumock GT. Economics considerations of Alzheimer's disease. Pharmacotherapy 1998;18:68–73.
[5] Alzheimer's Association. Alzheimer's disease Facts and Figures. Alzheimers Dement. 2014;10:e47–92.
[6] Yener GG, Erol B. Biomarkers in Alzheimer's disease with a special emphasis on event related oscillatory responses. Suppl. Clin. Neurophysiol. 2013;62:237–73. (Application of Brain Oscillations in Neuropsychiatric Diseases).
[7] Norouzi A, Mohd Shafry MR, Altameem A, Saba T, Rad AE, Rehman A, Uddin M. Medical image segmentation methods, algorithms, and applications. IETE Tech. Rev. 2014;31(3):199–213.
[8] Tanveer A, Taskeed J, Ui-Pil C. Facial expression recognition using local transitional pattern on Gabor filtered facial images. IETE Tech. Rev. 2013;30(13):47–52.
[9] Padilla P, López M, Górriz JM, Ramírez J, Salas-González D, Álvarez I. NMF-SVM based CAD tool applied to functional brain images for the diagnosis of Alzheimer's disease. IEEE Trans. Med. Imaging 2012;31(2):207–16.

[10] Yudong Z, Shuihua W, Zhengchao D. Classification of Alzheimer disease based on structural mag-netic resonance imaging by Kernel support vector machine decision tree. Prog. Electromagn. Res. 2014;144:171–84.

[11] Freeborough PA, Fox NC. MR image texture analysis applied to the diagnosis and tracking of Al-zheimer's disease. IEEE Trans. Med. Imaging 1998;17(3):475–9.

[12] Lene L, Lauge S, Akshay P, Erik BD, Mads N. Brain region's relative proximity as marker for Alzheimer's disease based on structural MRI. BMC Med. Imaging 2014;14(21):1–12.

[13] Rueda A, Gonzalez FA. Extracting salient brain patterns for imaging based classification of neurode-generative diseases. IEEE Trans. Med. Imaging 2014;33(6):1262–74.

[14] Illan IA, Górriz JM, Ramírez J, Meyer-Base A. Spatial component analysis of MRI data for Alzheimer's disease diagnosis: a Bayesian network approach. Front. Comput. Neurosci. 2014;156(8).

[15] Jing W, Zhilin Z. Identifying the neuroanatomical basis of cognitive impairment in Alzheimer's dis-ease by correlation and non-linearity – aware sparse Bayesian learning. IEEE Trans. Med. Imaging 2014;33(7):1475–87.

[16] P. Morgado, Automated diagnosis of Alzheimer's disease using PET images, MSc thesis at Electrical and Computer Engineering Dep., Higher technical institute, Technical University of Lisbon, September 2012.

[17] C. Ledig, R. Guerrero, T. Tong, K. Gray, A. Schmidt-Richberg, A. Makropoulos, R.A. Heckemann, D. Rueckert, Alzheimer's disease state classification using structural volumetry, cortical thickness and in-tensity features, in: MICCAI workshop Challenge on Computer-Aided Diagnosis of Dementia based on structural MRI data, 2014, pp. 55–64.

[18] Siqi L, Sidong L, Weidong C, Hangyu C, Pujol S, Kikinis R, Dagan F, Fulham MJ. Multimodal neu-roimaging feature learning for multiclass diagnosis of Alzheimer's disease. IEEE Trans. Biomed. Eng. 2015;62(4):1132–40.

[19] Kulkarni N, Bairagi V. Diagnosis of Alzheimer disease using EEG signals. Int. J. Eng. Res. Technol. 2014;3(4):1835–8.

[20] Blennow K, Hampel H, Weiner M, Zetterberg H. Cerebrospinal fluid and plasma biomarkers in Alzheimer disease. Nat. Rev. Neurol. 2010;6:131–44.

[21] Beuter A, Labric C, Vasilakos K. Transient dynamics in motor control of patients with Parkinson's disease. Chaos 1991;1:279–86.

[22] Glass HL. Nonlinear dynamics of physiological function and control. Chaos 1991;1:247–50.

[23] Sink KM, Holden KF, Yaffe K. Pharmacological treatment of neuropsychiatric symptoms of dementia: a review of the evidence. J. Am. Med. Assoc. 2005;293(5):596–608.

[24] Baker M, Akrofi K, Schiffer R, Michael W, Boyle O'. EEG patterns in mild cognitive impairment (MCI) patients. Open Neuroimage J. 2008;2:52–5.

[25] Besthorn C, Zerfass R, Geiger-Kabisch C, Sattel H, Daniel S, Schreiter-Gasser U, Frstl H. Discrimina-tion of Alzheimer's disease and normal aging by EEG data. Electroencephalogr. Clin. Neurophysiol. 1997;103(2):241–8.

[26] Van der Hiele K, Vein AA, Reijntjes RH, Westendorp RG, Bollen EL, van Buchem MA, van Dijk JG, Middelkoop HA. EEG correlates in the spectrum of cognitive decline. Clin. Neurophysiol. 2007;118(9):1931–9.

[27] Czigler B, Csikos D, Hidasi Z, Anna Gáal Z, Csibri E, Kiss E, Salacz P, Molnar M. Quantitative EEG in early Alzheimer's disease patients—power spectrum and complexity features. Int. J. Psychophysiol. 2008;68(1):75–80.

[28] Jeong J. EEG dynamics in patients with Alzheimer's disease. Artif. Intell. Med. 2004;115(7): 1490–505.

[29] Rene R, Francisco L, Alfredo A, Yuriem F, Lidice G, Yakeel Q, Maria AB. Spectral analysis of EEG in familial Alzheimer's disease with E280A presenilin-1 mutation gene. Int. J. Alzheimer's Dis. 2014;2014:180741.

[30] Dauwels J, Vialatte F, Cichocki A. Diagnosis of Alzheimer's disease from EEG Signals: where are we standing? Curr. Alzheimer Res. 2010;7(6):487–505.

[31] T. De Bock, S. Das, M. Mohsin, Early detection of Alzheimer's disease using nonlinear analysis of EEG via Tsallis entropy, in: IEEE Biomedical Sciences and Engineering Conference, May 2010, pp.1–4.

[32] F.J. Fraga, INRS-EMT, in: T.H. Falk, L.R. Trambaiolli, E.F. Oliveira, Towards an EEG-Based Biomarker for Alzheimer's Disease: Improving Amplitude Modulation, IEEE International Conference on Acoustics, Speech and Signal Processing (ICASSP), Univ. of Quebec, Montréal, QC, Canada, May 2013, pp. 1207–1211.

[33] Poil SS, Haan W, van der Flier WM, Mansvelder HD, Scheltens P, Linkenkaer-Hansen Klaus. Integrative EEG biomarkers predict progression to Alzheimer's disease at the MCI stage. Front. Aging Neurosci. 2013;5:58.

[34] Dauwels J, Srinivasan K, Reddy MR, Musha T, Vialatte F-B, Latchoumane C, Jeong J, Cichocki A. Slowing and loss of complexity in Alzheimer's EEG: two sides of the same coin? Int. J. Alzheimer's Dis. 2011;2011:539621. (Hindawi Access of Research).

[35] E. Gallego-Jutglà, M. Elgendi, F. Vialatte, J. Solé-Casals, A. Cichocki, C. Latchoumane, J. Jeong, J. Dauwels, Diagnosis of Alzheimer's disease from EEG by means of synchrony measures in optimized frequency bands, in: Conference proceedings: Annual International Conference of the IEEE Engineering in Medicine and Biology Society, IEEE Engineering in Medicine and Biology Society. Conference, Aug 28 2012–Sept 1, 2012, pp. 4266–4270.

[36] Cassani R, Falk TH, Fraga FJ, Kanda PA, Anghinah R. The effects of automated artifact removal algorithms on electroencephalography-based Alzheimer's disease diagnosis. Front. Aging Neurosci. 2014;6:1–13. 55.

[37] L.R. Traimboli, T.H. Falk, F.J. Fraga, R. Anghinah, A.C. Lorena, EEG sepectro-temporal modulation energy: a new feature for automated diagnosis of Alzheimer disease, in: Proc. Int. Conf. IEEE-EMBC, Boston, USA, 2011, pp. 3828–3831.

[38] T. Staudinger, R. Polikar, Analysis of complexity based EEG features for diagnosis of Alzheimer disease, in: Proc Intl. Conf. IEEE-EMBC, Boston, USA, 2011, pp. 2033–2036.

[39] Tsolaki A, Kazis D, Kompatsiaris I, Kosmidou V, Tsolaki M. Electroencephalogram and Alzheimer's disease: clinical and research approaches. Int. J. Alzheimer's Dis. 2014;2014:10. Article ID 349249.

[40] K. Akrofi, R. Pal, M.C. Baker, B.S. Nutter, R.W. Schiffer, Classification of Alzheimer's disease and mild cognitive impairment by pattern recognition of EEG power and coherence, in: 2010 IEEE International Conference on Acoustics Speech and Signal Processing (ICASSP), 14–19 March 2010, pp. 606–609 (doi: 10.1109/ICASSP.2010.5495193).

[41] Dauwels J, Vialatte F-B, Cichocki A. On the early diagnosis of Alzheimer's disease from EEG signals: a mini-review. Adv. Cogn. Neurodyn. 2010;2:709–16.

[42] Polikar R, Topalis A, Green D, Kounios J, Clark CM. Ensemble based data fusion for early diagnosis of Alzheimer's disease. Inf. Fusion 2008;9(1):83–95.

[43] Woon WL, Cichocki A, Vialatte F, Musha T. Techniques for early detection of Alzheimer's disease using spontaneous EEG recordings. Physiol. Meas. 2007;28(1):335–47.

[44] Ahmadlou M, Adeli H. Wavelet-synchronization methodology: a new approach for EEG based diagnosis of ADHD. Clin. EEG Neurosci. 2010;41(1):1–10.

[45] Cichocki A<ET-AL/>. EEG filtering based on blind source separation (BSS) for early detection of Alzheimer's disease. Clin. Neurophysiol. 2004;116(3):729–37.

[46] Dauwels J, Vialatte F. A Comparative study of synchrony measures for early diagnosis of Alzheimer's disease based on EEG. J. Neuroimage 2010;49(1):668–93.

[47] F. Vialatte, A. Cichocki, G. Dreyfus, T. Musha, S.L. Shishkin, R. Gervais, Early detection of Alzheimer's disease by blind source separation, time frequency representation and bump modeling, in: Artificial Neural Networks: Biological Inspirations – ICANN 2005, Lecture Notes in Computer Science, vol. 3696, 2005, pp. 683–692.

[48] Lehmann C, Koenig T, Jelic V, Prichep L, John RE, Wahlund LO. Application and comparison of classification algorithms for recognition of Alzheimer's disease in electrical brain activity (EEG). J. Neurosci. Methods 2007;161(2):342–50.

[49] G. Jacques, J. Frymiare, J. Kounios, C. Clark, R. Polikar, Multiresolution wavelet analysis and ensemble of classifiers for early diagnosis of Alzheimer's disease, in: Proc. of 30th IEEE Int. Conf. on Acoustics, Speech and Signal Proc (ICASSP 2005), vol. 5, Philadelphia, PA, March 2005, pp. 389–392.

[50] F. Vialatte, A. Cichocki, G. Dreyfus, T. Musha, T.M. Rutkowski, R. Gervais, Blind source separation and sparse bump modelling of time frequency representation of EEG signals: new tools for early detection

of Alzheimer's disease, in: 2005 IEEE Workshop on Machine Learning for Signal Processing, 28–28 Sept. 2005, pp. 27–32.

[51] Ghorbanian P, Devilbiss D. Identification of resting and active state EEG features of Alzheimer's disease using discrete wavelet transform. Ann. Biomed. Eng. 2013;41(6):1243–57.

[52] Kang Y, Escudero J, Shin D. Principal dynamic mode analysis of EEG data for assisting the diagnosis of Alzheimer's disease. IEEE J. Transl. Eng. Health Med. 2015;3:1–10.

[53] Trambaiolli LR, Lorena AC, Fraga FJ, Kanda PAM K, Nitrini R, Anghinah R. Does EEG montage influence Alzheimer's disease electroclinic diagnosis? Int. J. Alzheimer's Dis. 2011;2011:1–7. Article ID 761891.

CHAPTER TWO

Electroencephalogram and Its Use in Clinical Neuroscience

Electroencephalogram (EEG) was invented by German scientist Hans Berger in 1924. Thereafter, this recording technique has been one of the most used tools to study the brain activity and diagnose various neurological disorders. EEG is largely used due to its intrinsic simplicity and low cost as compared to that of the other recording techniques such as Magnetic Resonance Imaging (MRI), Single Photon Emission Computed Tomography (SPECT), or Positron Emission Tomography (PET) scans [1].

EEG records electrical fields generated in the brain. These electrical fields are generated by groups of pyramidal cells of neurons oriented perpendicular to the surface of the head. Ionic current flows are generated by neurons. In general, a group of neurons are modeled as a micro dipole when they produce synchronized electric fields. EEG records the electrical activity generated by the use of different micro dipoles in the surface of cortex regions. In approximate conditions, a group of 10^6 neurons orientated in the same direction with synchronized activity are enough to generate an electric field, which can be observable by the scalp. EEG is a complex combination of rhythms, recording the activity created in different parts at the same time. In general, EEG is present from before birth until brain death. The brain activity is related with any simple action that is performed by the human body such as movement of arms or focusing, shooting, reading, and many more. The study of the human brain activity results in improvement of our knowledge regarding the brain [2,3].

In previous studies, EEG readings were analyzed by visual interpretation and measures of EEG traces. Due to this, the results were unreliable. Similarly, due to the progress in the recent technology, data processing became efficient with computers; it became possible for the analysis of EEG signal digitally with parametric and nonparametric methods. Tremendous developments in electrical engineering and the fascination with the human brain have attracted researchers from different scientific fields to investigate EEG recordings. Due to this, a new subject called "EEG Signal Processing" came into existence. Today, EEG processing is highly used in medical technology due to the following reasons [1,4]:

1. EEG provides a high temporal resolution, which is not obtained by neuroimaging techniques. EEG provides a resolution of few milliseconds, whereas PET and fMRI are limited to seconds.
2. The knowledge of mechanisms that generates spontaneous EEG activity has increased.

EEG-Based Diagnosis of Alzheimer Disease
http://dx.doi.org/10.1016/B978-0-12-815392-5.00002-2

3. EEG recordings are costlier than the neuroimaging methods such as fMRI and PET scans.

4. New methods, that is, different signal processing methods, have recently emerged for EEG analysis such as Blind Source Separation (BSS) and time–frequency analysis such as Wavelet Analysis.

Due to the aforementioned reasons, EEG is regarded as a helpful tool in clinical neuroscience. The low cost of the EEG recording system is the reason that has resulted in its extended widespread use. Today, many doctors, neurologists, and neuroscientists are using EEG recording systems for diagnosing patients in clinical neuroscience studies. Some of these applications include the following:

• Monitoring alertness, coma, and brain death.
• Detecting neurodegenerative diseases such as Alzheimer's disease (AD), dementia, and Huntington's disease [5].
• Investigation of sleep disorders and epilepsy [1,5,19].
• For measuring the depth of anesthesia.
• Testing the drug effects [4].
• Serious games for e-learning and medical applications.

Although the EEG is largely used in the clinical neuroscience field, it is also progressively used in the other domains of study and applications. For example, EEG is also used in the Brain–Computer Interface (BCI) field to communicate the human brain with a machine. Different applications can be derived from BCI system, such as controlling video games or controlling a machine [6,7]. Other applications of EEG related to entertainment include "shippo" or "necomimi" projects, in which an EEG sensor is allegedly used to evaluate the mood of the subject and move a tail or cat's ear depending upon it [8,9].

2.1 EEG RECORDING AND MEASUREMENT

Acquisition of signals and images from the human body has become vital for early diagnosis of a variety of diseases. Such kind of data can be in the form of electrobiological signals such as an Electrocardiogram (ECG) from heart, Electromyogram (EMG) from muscles, EEG from brain, Magnetoencephalogram (MEG) from brain, Electrogastrogram (EGG) from stomach, and Electrooculogram (EOG) from eye nerves. Imaging techniques are also used for measurements such as Sonography (ultrasound image), Computerized Tomography (CT), MRI or fMRI, PET, and Single Photon Emission Tomography (SPET) [1,6].

The first electrical neural activities were registered using a simple galvanometer. The D'Arsonal galvanometer featured a mirror mounted on a movable coil. The light focused on mirror was reflected when a current passed the coil. Recent EEG systems consist of a number of dedicated electrodes, a set of different amplifiers (one for each of the channels), followed by filters and needle type registers. Later, more advances were made in EEG recording and measurement systems. Hence, to analyze EEG signals, it was digitized and stored in some form. This required sampling quantization and encoding of the signal. The computerized systems allow variable settings, stimulations, and sample frequency, and some are equipped with simple or advanced signal processing tools for preprocessing the signals [1,10].

The format of reading the EEG data is different for different EEG machines. These formats are easily convertible to spreadsheets, which are easily readable by signal processing software packages such as MATLAB and EEGLAB. The EEG recording electrodes and their proper function are important for acquiring high-quality EEG data. In EEG recording, different types of electrodes are often used such as the following:

- Disposable (gel-less and pregelled types),
- Reusable disc electrodes (gold, silver, stainless steel, tin, etc.),
- Head bands and electrode caps,
- Saline-based electrodes, and
- Needle electrodes.

EEG is recorded though different sensors (electrodes) placed on the scalp. Sensors are disks of 5 mm, generally made up of Ag/AgCL (silver/silver chloride). For the EEG recordings, different electrode placement systems have been proposed such as International 10–20 system, Maudsley system, and 10–10 system. Nowadays, the most commonly used system is the International 10–20 electrode placement system. The International 10–20 system of electrodes placement presents a uniform coverage of the entire scalp. The International Federation of Societies for Electroencephalography and Clinical Neurophysiology has recommended the conventional electrode setting, which is termed as 10–20 electrode placement system for 21 electrodes as shown in Fig. 2.1. This system is based on an iterative subdivision of arcs on the scalp starting from the following craniometric reference points: Nasion (Ns), Inion (In), Left (PAL), and Right (PAR) preauricular points [6,7]. The standard set of electrodes is detailed in Figure, showing the location of 21 recording electrodes. The 10 and 20 indicate that the distance between adjacent electrodes is either 10% or 20% of a specified distance measured using specific anatomical landmarks, for example, the total distance between the front and back or left and right of the head. Electrodes are numerated with a letter and a subscript. The letter specifies the anatomical area where the electrode corresponds to the following: prefrontal or frontopolar (Fp), frontal (F), central (C), parietal (P), occipital (O), temporal (T), and auricular (A). The subscript is the letter z, indicating zero or midline placement, or a number indicating lateral placement. Electrodes with even numbers are placed on the right side of the head, while odd numbers correspond to the left side of the brain. The number of positioning electrodes increases with increasing distance from the anterior posterior midline

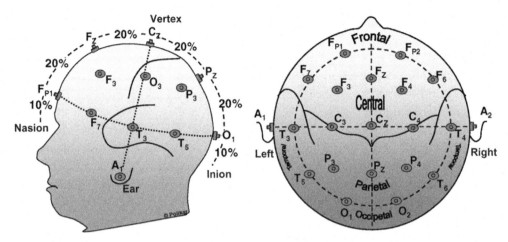

Figure 2.1 *International 10–20 electrode placement of EEG signals [11].*

of the head. Once the electrodes are placed, different montages can be used for the recording of brain electrical potentials [6,10] (Fig. 2.1).

EEG can be recorded using referential or bipolar montages. In referential montages, the voltage differences between all electrodes and a common electrode (reference) are recorded. In bipolar montages, however, instead of using a reference electrode, the voltage difference between two designated electrodes is recorded (i.e., each electrode pair is considered as a channel). A major disadvantage of the referential montage is that there is no single reference electrode optimal for all situations, since no reference is truly inactive. A bipolar montage reduces the effects of common noise/artifacts and eliminates the influence of contaminated references.

2.2 EEG RHYTHMS

EEG signals are highly sensitive of subjects and states; hence, EEG rhythms change depending on the subject task. But, generally five major types of continuous rhythmic EEG activities are recognized in the recordings [11,12]. They are divided into different frequency bands.

The typical EEG frequency and their respective frequency bands are classified as follows:
- δ rhythms, found in the frequency bands of 1–4 Hz (δ band)
- θ rhythms, found in the frequency bands of 4–8 Hz (θ band)
- α rhythms, found in the frequency bands of 8–13 Hz (α band)
- β rhythms, found in the frequency bands of 13–30 Hz (β band)
- ψ rhythms, found in the frequency bands of 30–100 Hz (ψ band)

A brief description of these EEG rhythms is presented in the following paragraphs:

- **Delta Rhythms (δ: 0.5–4 Hz)**

  They are the slowest of all rhythms. But, they present higher amplitude of all. These rhythms are observed when the subject is sleeping [1]. The higher value is observed when the subject is in deep sleep (Fig. 2.2).

- **Theta Rhythms (θ: 4–8 Hz)**

  The term theta is chosen to allude to its presumed thalamic origin. They are mainly associated with drowsiness, childhood, adolescence, and young adulthood. Theta (θ) waves are also associated with access to unconscious material, creative inspiration, and deep meditation [1,4,5].

  These are also found during problem solving; for example, while solving mathematical problems such as adding and subtracting. It is located in the prefrontal part of the cortex. The changes in the rhythms of theta waves are examined for maturational and emotional studies.

- **Alpha Rhythms (α: 8–12 Hz)**

  They are associated with relaxed states. They appear in the posterior half of the head and are commonly found over the occipital region of the brain. They commonly appear as the round- or sinusoidal-shaped signal. Hence, alpha waves have been thought to indicate both a relaxed awareness without any attention or concentration [1,4]. Different alpha rhythms are found in the human cortex: μ rhythm, alpha occipital rhythm, and alpha parietal rhythms. Let us discuss the aforementioned alpha rhythms in brief (Fig. 2.3).

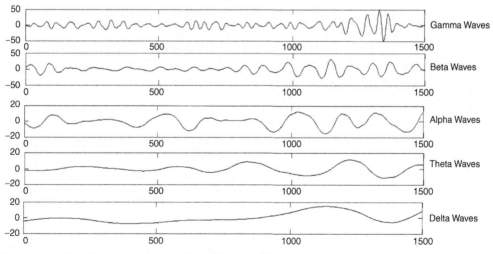

Figure 2.2 *Classification of EEG signal in different sub-bands.*

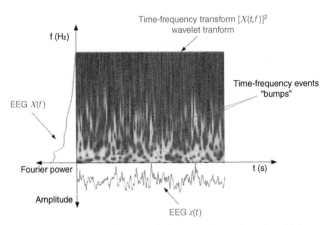

Figure 2.3 *Slowing of the EEG signals: EEG signal, shown in time-domain x(t), frequency domain X(f), and time–frequency domain |X(t,f)| [5].*

- μ Rhythms have widely been used in implementation of the BCI applications. They have this name due to the similarity that these rhythms present with μ letter (with sharp −ve peaks). They are found in the frequency range close to 10 Hz. The change of activity is a well-known phenomenon due to the synchronization of the groups of neurons in the motor cortex region. When the limbs are inactive, the μ rhythms presents activity, whereas when a subject moves his limbs, the rhythms present a decrease of amplitude, which is termed as α desynchronization. These properties are used in BCI systems to detect motor imagination.
- α Wave has higher amplitude over the occipital areas and has amplitude of normally less than 50 μV. The origin and biophysiological significance of an alpha wave is unknown, and more research is ongoing to understand how this phenomenon originates from cortical cells of brain regions.

- **Beta Rhythms (β: 12–30 Hz)**
  Beta wave is a low-amplitude signal with multiple and varying frequencies of the brain associated with active thinking, active attentions, focus on the outside world, or solving concrete problems; it is normally found in adults. A high-level beta wave is observed when the human is in a panic state. Basically, there are different types of beta rhythms such as Beta Rolandic rhythms and Beta Frontal rhythms [1].
  - Beta (β) Rolandic rhythms are observed as spontaneous activity recorded on the electrodes located close to the sensor motor area (C3 and Cz).
  - A Beta (β) Frontal rhythm appears during cognitive tasks related with decision-making. Maximum activity is observed in F3, F4, and Fz electrodes.

- **Gamma Rhythms (ψ: 30–100 Hz)**
  Initially, gamma (ψ) rhythms were not studied, since previous EEG recording systems were unable to record signals above 25 Hz. The first article describing these

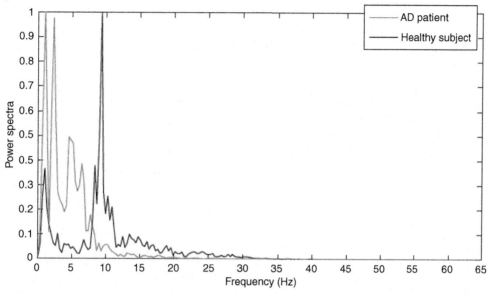

Figure 2.4 *Comparison between a mild AD patient and a healthy subject using power spectrum.* The graph is presented for Fz electrode [12].

rhythms appeared in 1964. Gamma rhythms appear in the higher mental activity, including perceptions, fear, etc. [1,4]. The gamma wave band is proved to be a good indication of event-related synchronization (ERS) of brain and is used to demonstrate the locus for right and left index finger movement, bilateral area for tongue movement, etc. (Fig. 2.4).

Waves in frequencies higher than the normal activity range of EEG, specifically in the range of 200–300 Hz, are found in cerebella structures of animals, but they do not play any role in clinical neurophysiology. EEG signals are the projections of neural activities or subjects that are attenuated by leptomeninges, cerebrospinal fluid (CSF), dura mater, bone, and scalp. Cartographic discharges show amplitude readings over 0.5–1.5 mV for spikes. But on the scalp, the amplitude lies within the range of 10–100 μV.

2.3 EARLY DIAGNOSIS OF ALZHEIMER'S DISEASE BY MEANS OF EEG SIGNALS

In previous sections, we discussed the recordings techniques known as EEG, and in the previous chapter, we presented the information about AD. This section describes the state-of-the-art of using EEG as an early diagnosis tool to diagnose AD in its early stage. AD diagnosis is done by using a combination of different tools. An early diagnosis of AD may benefit the patient in different ways, such as it gives the time to inform them about the disease, taking economical depositions to plan for the future needs of the patients to take the symptoms delaying medications which are effective in the first stage of disease [5]. Study involving early diagnosis of AD by means of EEG helps doctors to diagnose subjects suffering from Alzheimer's. It provides extra information in combination with laboratory tests and physical and neurological examinations.

In recent research findings, it was observed that many different groups had focused on AD diagnosis using EEG. There are several arguments supporting this research direction. It is observed that AD causes cortical dementia through which damages are induced in brain cells, which are reflected in the EEG recordings. Another reason in this contradicts to easy recording of EEG signal in the clinical environment. Many a times, the EEG recordings are done when the cortical subjects are in a resting state with eyes closed. This procedure benefits the elderly. As there is no required stimulation device and there are no cognitive tasks to perform, subjects are less fatigued and anxious to perform the tasks [14]. EEG recordings in the resting state can be recorded in high comparable experimental conditions for all subjects with MCI, patients with mild AD, and healthy subjects.

Three different major effects are observed in EEG recordings of AD patients. Slowing of EEG in AD patients, enhanced complexity of EEG signals, and perturbations in synchrony measures are observed. Even though these effects are repeated reported in literature, many a times, they are not detectable due to the high variability in AD patients. Hence, any of the observed effect is used to facilitate the reliability of diagnosing AD in early stages.

2.3.1 Slowing of EEG signals in AD patients

One of the major effects is EEG "slowing"; many studies have shown that Mild Cognitive Impairment (MCI) and AD cause EEG signals to slow down. MCI/AD is associated with an increase of power in low frequencies (delta and theta bands, 0.5–8 Hz) and a decrease of power in higher frequencies (alpha and beta bands, 8–30 Hz) [1,5]. However, increased gamma band power (30–100 Hz) has been reported in MCI/AD patients compared to healthy age-matched control subjects. Time–frequency maps of EEG signals are often sparse, as can be seen from most energy is contained in specific regions of the time–frequency map (bumps), corresponding to transient oscillations [14,15]. A procedure was proposed to extract such transient oscillations from time–frequency maps; it was shown that transient oscillations in the EEG of MCI and AD patients occur more often at low frequencies compared to healthy control subjects. In other words, those transient oscillations also exhibit slowing.

To facilitate the comparison of power in the frequency domain, the Fourier transform (DFT) is applied. A comparison between the power spectra of an AD patient and a healthy subject is also presented. This figure illustrates the clear difference existing between the power spectra of a healthy subject (red in the figure) in comparison with the power spectra of an AD patient (blue in the figure). In this figure, the differences between the two power spectra are clearly presented. Healthy subject has a peak of power between 9 and 10 Hz (called α band), whereas the AD patient has different peaks of power below 8 Hz, standing for the δ band, being the δ band that presents the higher values of power spectra. Studies have mainly used the differences of power in the different frequency bands (δ, θ, α, and β bands) to differentiate how AD patients and healthy subjects are classified.

2.3.2 Perturbations in EEG synchrony

One of the most important features of the brain is not the high number of neurons that it contains but the abundant connectivity between these. This high connectivity is observed in synchronous activity, as neurons in anatomically connected structures tend to fire synchronously. Many different measures of synchrony have been employed in the physical sciences, signal processing, and in the study of neurobiology. Many of these techniques have overlapping properties, while some are substantially different. All seek to quantify the relationship between two signals or sensors and, by extension, the sources they represent. Numerous studies have reported decreased EEG synchrony in MCI/AD patients [15]. The main problem is that most studies use just one measure or very few measures, and many of those studies analyze different data sets; consequently, it is difficult to rigorously compare the various measures. One should keep in mind, however, that it is hard to directly interpret results obtained with synchrony measures. Synchrony measures obtained from EEG signals may be significantly affected by brain events other than changes of synchrony, and by choices (like the reference electrodes) that necessarily have to be made during the analysis. Several of the synchrony measures can be applied such as the Pearson Correlation Coefficient, Magnitude and Phase Coherence, Granger Causality, and Phase Synchrony [11,12]. A variety of measures have been used in the literature in order to compute the synchrony of EEG recordings. Some typical measures include Coherence, Granger Measures, State Space-based Synchrony measures, Phase Synchrony (PS), and stochastic event synchrony measures. All these measures seek to quantify the relationships between two or more signals.

2.3.3 Reduced complexity in EEG signals

Nonlinear Dynamical Analysis (NDA) of the EEG has proved the decreased complexity of EEG patterns and reduced functional connections in AD. NDA has offered valuable information on cortical dynamics when applied to EEG, due to the high complexity of biological signals. Several studies have investigated the complexity of EEG signals in MCI and AD patients. Various complexity measures have been used to quantify EEG complexity; some of them stem from information theory such as the Tsallis entropy [15], approximate entropy [15], multi-scale entropy [16,17], sample entropy and mutual information, and Lempel–Ziv complexity [18]. All studies observed that the EEG of MCI and AD patients seems to be more regular (and, equivalently, less complex) than of age-matched control subjects. Due to the MCI/AD-induced loss of neurons and perturbed anatomical and/or functional coupling, fewer neurons interact with each other, and the neural activity patterns and dynamics become simpler and more predictable. Moreover, it is possible that the MCI/AD-induced phenomenon of reduced complexity is related to slowing, since "slower" (low-pass) signals are intrinsically more regular.

Results presented in Dauwels et al. [5] evaluated the use of complexity-based measures together with the effect of slowing of EEG as already discussed in the previous section. Results show that, for the data sets used, complexity measures are highly correlated with measures used to compute the effect of slowing of EEG.

SUMMARY

In this chapter, we present an overlook of EEG system in medical technology. EEG signal is used to record the electrical activity of brain regions. EEG signal is a nonstationary signal, and it makes analysis and interpretation much difficult. EEG signal is classified into five sub-bands as per various frequency ranges, and each sub-band has its own characteristics. The chapter also discusses the state-of-the-art of recording and measuring of EEG signal. Section 2.3 highlights the use of EEG signals in diagnosis of AD. It is discussed that certain abnormalities are observed in the EEG signals of Alzheimer's patients. Slowing of EEG signals, reduced complexity, and perturbations in EEG synchrony measures are some of the abnormalities observed. In preceding chapters, the verification of these measures is explored briefly for Alzheimer's disease diagnosis.

REFERENCES

[1] Nunez PL, Srinivasan. Electric fields of the brain: the neurophysics of EEG. 2nd ed. Oxford University Press; 2006.
[2] Illan IA, Górriz JM, Ramírez J, Meyer-Base A. Spatial component analysis of MRI data for Alzheimer's disease diagnosis: a Bayesian network approach. Front Comput Neurosci 2014;156(8).
[3] Morgado P. Automated diagnosis of Alzheimer's disease using PET images [M.Sc. thesis]. Electrical and Computer Engineering Department, Higher Technical Institute, Technical University of Lisbon; 2012.
[4] Kropotov JD. Quantitative EEG, event-related potentials and neurotherapy. Elsevier Inc.; 2009.
[5] Dauwels J, Vialatte F, Cichocki A. Diagnosis of Alzheimer's disease from EEG signals: where are we standing? Curr Alzheimer Res 2010;7(6):487–505.
[6] Schalk G, Mellinger J. A practical guide to brain computer interfacing with BCI2000. Springer Science and Business Media; 2010.
[7] Fisch BJ. Fisch and Spehlmann's, EEG primer: basic principles of digital and analog EEG. Third revised and enlarged edition Elsevier; 1999.
[8] Neurowear. Neocomimim, a brainwave sensor to control cat's ear. Available from: http://neurowear.com/projects_detail/necomimi.html; 2014.
[9] Neurowear. Shippo, a brainwave sensor to control a tail. Available from: http://neurowear.com/projects_detail/shippo.html; 2014.
[10] Graimann B, Allison BZ, Pfurtscheller G. Brain-computer interfaces: revolutionizing human-computer interaction. Springer; 2011.

[11] Polikar R, Topalis A, Parikh D, Green D, Frymiare J, Kounios J, Clark CM. An ensemble based data fusion approach for early diagnosis of Alzheimer's disease. Inf Fusion 2008;9(1):83–95. Elsevier.

[12] Gallego Jutgla E. New signal processing and machine learning methods for EEG data analysis of patients with Alzheimer's disease [Ph.D. thesis]. Department of Information Technology, University of Central De Catalunya; 2014.

[13] Gallego-Jutglà E, Elgendi M, Vialatte F, Solé-Casals J, Cichocki A, Latchoumane C, Jeong J, Dauwels J. Diagnosis of Alzheimer's disease from EEG by means of synchrony measures in optimized frequency bands. In: Conference proceedings: Annual International Conference of the IEEE Engineering in Medicine and Biology Society; August 28–September 1 2012. p. 4266–4270.

[14] Vialatte F, Cichocki A, Dreyfus G, Musha T, Rutkowski TM, Gervais R. Blind source separation and sparse bump modelling of time frequency representation of EEG signals: new tools for early detection of Alzheimer's disease, In: 2005 IEEE Workshop on Machine Learning for Signal Processing; 28 September 2005. p. 27, 32.

[15] Cover TM, Thomas JA. Elements of information theory. New York: Wiley; 1991.

[16] Shannon CE. A mathematical theory of communication. Bell Syst Tech J 1948;27:379–423. 623–656.

[17] Ziv J, Lempel A. A universal algorithm for sequential data compression. IEEE Trans Inf Theory 1987;23:337–43.

[18] Lempel A, Ziv J. On the complexity of finite sequences. IEEE Trans Inf Theory 1976;22:75–81.

[19] Harpale VK, Bairagi VK. Time and frequency domain analysis of EEG signals for seizure detection: a review. In: IEEE International Conference Microelectronics, Computing and Communications (MicroCom) 2016, NIT Durgapur; January 2016, p. 1–6.

CHAPTER THREE

Role of Different Features in Diagnosis of Alzheimer Disease

3.1 INTRODUCTION

There is an immense need for the development of new methods and techniques for early identification of Alzheimer's disease, and in this regard a number of brain imaging techniques facilitate providing noninvasive ways for visualization of brain atrophy. The earlier diagnosis of any disease is not only challenging but also crucial for further treatment [1,2]. The detailed block diagram and detailed approach used in our research work is shown in Figs. 3.1 and 3.2.

Let us discuss each of the block in more detail.

3.1.1 Phase 1: Preprocessing

Any signal processing algorithm like segmentation and feature extraction relies significantly on the quality of the signal obtained. The quality of electroencephalogram (EEG) signal deteriorates either during its acquisition process or afterwards. The process of EEG acquisition may incorporate certain artifacts. For example, noise problem and intensity in homogeneity. These artifacts occur due to the power lines interferences in the machine. Some biological artifacts are also observed in the recordings of the EEG data. They may occur due to the human efforts. Eye blinking of patients, muscle activity, and variations in EEG electrodes are some of the causes for signal contamination in EEG recordings. Hence to remove these and to optimize the signal for further analysis and evaluation, a number of preprocessing steps are required. Signal preprocessing can significantly increase the visual reliability of the signal. It involves a set of techniques that enhances or eradicates certain details of the signals in order to efficiently process it for further analysis. First, we have used the bandpass filter for filtering the EEG signal from 0.5 to 30 Hz. In our proposed approach the set of preprocessing techniques includes Independent Component Analysis and wavelet-based denoising for eliminating the unwanted data. The details of these steps are given explained in further sections [1,3].

3.1.2 Phases 2 and 3: Segmentation and feature extraction

Biosignals processing serves best in early detection of several diseases, such as Alzheimer's disease as they hold necessary information to distinguish healthy controls and Alzheimer's disease patients. But, the major concern here is the huge data size of EEG signals.

EEG-Based Diagnosis of Alzheimer Disease
http://dx.doi.org/10.1016/B978-0-12-815392-5.00003-4

37

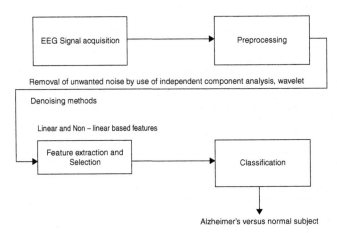

Figure 3.1 *Basic system approach.*

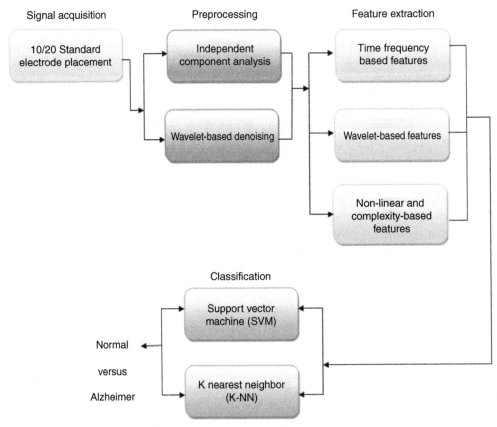

Figure 3.2 *Detailed working approach of the system.*

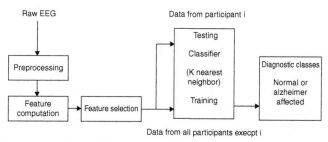

Figure 3.3 *System approach used for diagnosis of Alzheimer disease.*

In order to classify these signals by the classifier, the computational time is enormous. Moreover not all information in the signal is required for the classification purposes as most of the information is irrelevant. For this purpose feature extraction is performed to find more relevant and discriminative features [1], in order to classify signals more efficiently. In recent literature a vast variety of features has been extracted from EEG signals for the identification of AD. These features are time–frequency–based features, wavelet-based features and complexity-based features. These selected features are affected in the earliest stages of AD. Hence the contribution of presented research is toward identification of AD patients in early stages, using a smaller feature set which results in lower computational expense.

3.1.3 Phase 4: Classification

Different types of classifiers, both supervised, as well as unsupervised are efficiently used in machine learning and pattern recognition. In our proposed system, we have used supervised approach for classification between two classes, that is, support vector machine and K nearest neighbor classifier. In addition to these classifiers, we can also use ensemble of these classifiers to verify the enhanced accuracy rate. Fig. 3.3 shows the system approach of the study. Let us discuss each step used in detail.

3.2 WHAT IS FEATURE EXTRACTION?

Feature extraction means process of analyzing digital signals to distinguish pertinent signal characteristics (i.e., signal features related to the person's intent) from extraneous content and representing them in a compact form suitable for translation into output commands [2]. It is necessary to have strong correlations with the user's intent for calculated features. Since, much of the relevant (i.e., most strongly correlated) brain activities are either transient or oscillatory. The most commonly extracted signal features in proposed system are time-triggered EEG or electrocorticography (ECoG) response amplitudes and latencies, power within specific EEG or ECoG frequency bands, or firing rates of individual cortical neurons. Environmental artifacts and physiologic artifacts, such as electromyographic (EMG) signals are avoided or either removed to ensure accurate measurement of the brain signal features.

3.3 NEED OF FEATURE EXTRACTION

Extraction of relevant and appropriate feature is one of the most critical and significant procedures for classification of data between two groups. As feature extraction process has a direct impact on classification performance of the system. If extracted features are not expressive, it can lead to unsatisfactory classification performance. In these cases, classification process may be optimal but inadequate features may not provide satisfactory classification results for the given problem [4,5]. Feature extraction is process that is generally based on the reduction of the existing features by applying a several transformations to obtain a lower dimensional space which can better represent the target concept. In process of feature extraction, features are extracted from the preprocessed and digitized EEG signal. In the easiest form, a certain frequency range is selected and the amplitude relative to some reference level is measured and processed. For instance, computed features are certain frequency bands of a power spectrum. The power spectrum (which characterizes the frequency content of the EEG signal) can be computed using fast Fourier transform (FFT) or any other tool, the transfer function of an autoregressive model or wavelet transform. On the other hand, if the feature sets are specific, any classifier can classify them. In present research, various different features are explored for distinguishing between two groups, that is, Alzheimer's disease patients and normal subjects. In this section, we will be discussing the different features used in the study for classifying the data in two groups. We have involved the use of time–frequency based features, wavelet-based features and complexity-based features in our study [3,6].

3.4 LINEAR FEATURES

In this part, different features for classifying the Alzheimer's disease and healthy patients are discussed. The following features are used for feature extraction process because these features gives better results in terms of classification accuracy as discussed in various research articles.

The following features are discussed briefly:
1. Spectral features
2. Wavelet-based features
3. Complexity-based features

3.4.1 Spectral features

This feature includes the variety of frequency-based features used in the study. First, the raw EEG signal is decomposed into the different EEG sub bands, such as

Delta-, Theta-, Alpha-, Beta-, and Gamma-band by means of wavelet decomposition. Wavelet-based transformation of biosignals, such as EEG signal to represent in power distribution over both time and frequency continues to be a promising approach and it is suitable for spectral analysis for detection of brain diseases as compared to FFT. Basically, there are two types of wavelet analysis: (i) continuous wavelet transform (CWT) and (ii) discrete wavelet transform (DWT). Both the DWT and CWT are used in EEG classification and analysis as discussed in literature. Statistical features are calculated using various wavelet decomposition levels. DWT is computationally efficient as compared to CWT and it is used widely than CWT. In this method, DWT is used for wavelet decomposition. DWT analyzes the EEG signal of different temporal resolutions through its decomposition into different frequency bands by utilizing a scaling and a wavelet function associated with low- and high-pass filters [3,5–7].

The original EEG signal $x(t)$ basically forms the discrete time signal $x[i]$, which is first passed through a half-band high-pass filter $g[i]$, and a low pass filter $h[i]$. The filtering followed by subsampling constitutes the one level of decomposition and it is expressed as:

$$d[k] = \sum_n x[i].g[2k-i] \tag{3.1}$$

$$a[k] = \sum_n x[i].h[2k-i] \tag{3.2}$$

where d[k] and a[k] are first level detailed and approximate coefficients at translation k, which are the outputs of the high-pass and low-pass filters after the subsampling process. The same procedure is repeated further for further decomposition as many times as desired or until no more subsampling is possible. At every level of decomposition, it results in half the time resolution (due to subsampling) and double the frequency resolution, allowing the signal to be analyzed at different frequency ranges with different resolutions.

From study of different families of mother wavelets, Daubechies family possesses the number of characteristics that are ideal and suitable for EEG analysis, including (1) the well understood and smoothing characteristics of Daubechies, (2) detection of changes in EEG important for detecting Epileptic activity. In present research, different mother wavelets from Daubechies families were tested. But, "db8" wavelet was used for decomposition of EEG signal. Five levels of decomposition are performed resulting in D_1 (approximately related to the gamma spectral frequency band) through D_5 (approximately related to the upper delta spectral frequency band) and A_1 through A_5 (approximately related to lower delta spectral frequency band), as shown in Fig. 3.4. This shows the extract frequency ranges and their corresponding EEG major spectral frequency bands [3,6].

Table 3.1 gives an overview of the different EEG sub bands after obtaining the wavelet decomposition at level 5.

Using the above frequency bands obtained after classification of EEG signals in several different bands, such as the Delta, Theta, Alpha, Beta, and Gamma; relative power features from each bands is computed to observe the changes in power between two

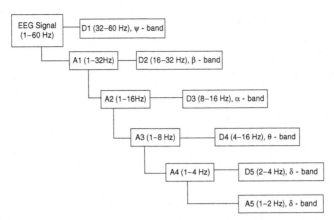

Figure 3.4 *Five levels decomposition of an electroencephalogram (EEG) signal; D1–D5 and A5 are the DWT representation of the signal.*

Table 3.1 DWT sub-band frequencies and the corresponding approximate major brain frequency bands

Sub-band	Frequency range	Corresponding EEG frequency band (Hz)
D_1	30–60	Ψ (>30)
D_2	15–30	β $(13–30)$
D_3	7.5–15	α $(8–13)$
D_4	3.75–7.5	θ $(4–8)$
D_5	1.875–3.75	δ $(2–4)$
A_5	1–1.875	δ $(1–2)$

groups. Several research findings have shown the changes in the EEG power spectra due to AD. It consist of increase in the delta and theta band powers, together with a decrease in alpha and beta band powers, thus suggesting a slowing of the EEG signal. Spectral power features measures the power present in each of the five conventional EEG frequency bands, namely: 0.1–4 Hz (delta), 4–8 Hz (theta), 8–12 Hz (alpha), 13–30 Hz (Beta), and 30–100 Hz (gamma).

The judgment of spectral characteristics of the EEG activity is basically based on the power spectral density (PSD) of each EEG epoch, which is computed as the Fourier transform of its autocorrelation function [8]. The PSD is normalized by the total power in the considered broadband (1–40 Hz) to obtain a normalized PSD (PSD_n):

$$PSD_n(f) = \frac{PSD(f)}{\sum_{f=1HZ}^{40Hz} PSD(f)} \tag{3.3}$$

So that:

$$\sum_{f=1\text{Hz}}^{40\text{Hz}} \text{PSD}_n\left(f\right) = 1 \qquad\qquad (3.4)$$

If f_{low} and f_{high} are low- and high-cutoff frequencies of each band (e.g., f_{low} = 1 Hz and f_{high} = 4 Hz for Delta band, then Relative Power (RP) is calculated as,

$$\text{RP} = \sum_{f\text{low}}^{f\text{high}} \text{PSD}_n\left(f\right) \qquad\qquad (3.5)$$

By using the above listed formula's and use of discrete Fourier transform, we calculated the spectral-based features in each bands of EEG signals of both the groups.

3.4.2 Wavelet-based features

Fourier transforms is used extensively for analyzing various signals, such as biomedical signals like electrocardiogram (ECG), EEG, speech signals, and many more. Many of the times, Fourier transform is not always the correct or appropriate method for analysis of nonstationary signals since its spectral content changes with time. A Fourier signal analyzes spectral components residing in the signal. On the other hand, it does not provide information related to the time localizations of the spectral components. A time–frequency representation provides time-localizations of the spectral components of the signal which provides necessary significant information. To represent power distribution of EEG signals over both time and frequency, wavelet transform proves to be a promising approach and it is suitable for spectral analysis for detection of brain disorders, such as Alzheimer's, Epilepsy, tumors, and many more as compared to FFT [3–6].

In previous decomposition of signal by means of wavelet technique, we have decomposed the signal at five levels for separating the signal in different bands of frequency. The wavelet coefficient of the decomposed signal is very large and it is not suitable to use directly for pattern recognition process. Therefore, features are extracted to reduce the signal to its representation set of feature vectors by simplifying the description of a large set of data [9].

The features can be extracted in time, as well as frequency domain. Statistical approach of the time domain features proves to be most simple and commonly used feature to represent the large sets of data. Statistical features, such as mean, mode, standard deviation, and variance can be used in general. In present research work, mean power of the wavelet coefficients was computed using following formula:

$$\text{Mean} = \frac{1}{n}\sum_{i=0}^{n-1} x_i^2, j = 1...N \qquad\qquad (3.6)$$

where x_i's are the computed coefficients of the signal at each subband, n denotes number of coefficients at each band and N denotes number of band. These features values were computed at each level of DWT decomposition separately for each recording block from each task of each subject [9].

Variance of the wavelet coefficient discrete-time series is computed using (7), where n denotes length of the discrete-time signal and x represents the level of the signal for each n.

$$\sigma^2 = \frac{1}{n-1}\sum_{i=0}^{n-1}(x_i - \mu)^2 \tag{3.7}$$

3.4.3 Complexity based features/nonlinear features

The following complexity based features are used for feature extraction and classification of EEG data in two groups. The significance of all these features is discussed briefly in next chapter. Let us discuss in short about these features.

1. Spectral entropy

   Spectral entropy indicates the amount of unpredictability and disorder in spectrum of EEG. Higher complexity is achieved if higher amount of spectral is entropy is observed [9,10,11]. It is computed in following manner:

   a. For the given signal $x(t)$, compute $S(f)$, the PSD, as the Fourier transform of the autocorrelation function of the signal $x(t)$.

   b. Depending upon the frequency of interest; extract the Power in the spectral band from 0.5–30 Hz.

   c. After calculation of the spectral band power, normalize the power in the given band of interest.

   d. Calculate the spectral entropy by using formula,

   $$SE = \sum_{f=0.5}^{40} S(f) \star \ln\frac{1}{s(f)} \tag{3.8}$$

2. Spectral centroid

   Spectral centroid (SC) measures the shape of the spectrum of EEG signals. A higher value of SC corresponds to more energy of the signal being concentrated within higher frequencies. Basically, it measures the spectral shape and position of the spectrum [9,10,11].

   It is computed as follows:

   a. Let $x_i(n), n = 0,1,...N–1$ be the sample of the ith frame, with $X_i(k), k = 0,1,...N–1$ as the discrete Fourier transform (DFT) coefficients of the sequence.

   b. Compute the SC of the each frame as:

   $$C(i) = \frac{\sum_{k=0}^{N-1} k|Xi(k)|}{\sum_{k=0}^{N-1}|Xi(k)|} \tag{3.9}$$

The mean value of the SC across all the frames can be used as the SC feature for each epoch of the frame of the EEG signal.

3. Spectral roll-off

   Spectral roll-off represents the frequency below which a certain percentage (usually 80%–90%) of the magnitude distribution of the spectrum is concentrated in the spectrum. It is computed as follows:

   a. Let $x_i(n)$, $n = 0,1,...N-1$ be the sample of the ith frame, with as the discrete $X_i(k)$, $k = 0,1,...N-1$ Fourier transform (DFT) coefficients of the sequence.

   b. Compute the spectral roll-off as the sample that satisfies,

$$\sum_{k=0}^{k(i)} |X_i(k)| = \frac{P}{100} \sum_{k=0}^{N-1} |X_i(k)| \tag{3.10}$$

Where, the P parameter is sometimes chosen between 80 and 100.

4. Zero crossing rate

   Zero crossing rate of any signal frame is the amount at which a signal changes its sign during the frame. It denotes the number of times the signal changes value, from positive to negative and vice versa, divided by the total length of the frame.

   Zero crossing rate is computed as:

   a. Let $x_i(n)$, $n = 0,1,...N-1$ be the samples of the ith frame.

   b. Zero crossing rate of each frame is calculated as:

$$Z(i) = \frac{1}{2N} \sum_{n=0}^{N-1} |sgn[x_i(n)] - sgn[x_i(n-1)]| \tag{3.11}$$

where,

$$sgn[x_i(n)] = \begin{cases} 1, x_i(n) \geq 0 \\ -1, x_i(n) < 0. \end{cases}$$

3.5 CONCLUSIONS

In this chapter, need of features extraction is discussed briefly. Along with this, several features, such as linear and nonlinear features are explored briefly. The chapter gives us overview of various features used for classifying the data between two groups. Linear features, such as spectral and wavelet features are discussed considering its physical significance and nonlinear features, such as spectral entropy, centroid, roll-off, and zero crossing rates are explored with its physical significance on computation of EEG signals. Thus, this chapter gives information about various features used for analyzing of EEG signals for distinguishing between Alzheimer's patient and normal subjects.

REFERENCES

[1] Saima F, Muhammad AF, Huma T. An ensemble-of-classifiers based approach for early diagnosis of Alzheimer's disease: classification using structural features of brain images. Comput. Math. Methods Med. 2014;2014:11. Article ID 862307.

[2] Shih JJ, Krusienski DJ, Wolpaw JR. Brain-computer interfaces in medicine. Mayo Clin. Proc. 2012;87(3):268–79.

[3] Sanei S, Chambers JA. EEG Signal Processing. Chichester: John Wiley & Sons; 2007.

[4] Amin HU, Malik AS, Ahmad RF, et al. . Australas. Phys. Eng. Sci. Med. 2015;38:139.

[5] Polikar R, Keinert F, Greer MH. Wavelet analysis of event related potentials for early diagnosis of Alzheimer's disease. In: Petrosian AA, Meyer FG, editors. Wavelets in Signal and Image Analysis. Computational Imaging and Vision, vol. 19. Dordrecht: Springer; 2001.

[6] Daubechies I. Ten Lectures on Wavelets. Philadelphia, PA: Society for Industrial and Applied Mathematics; 1992.

[7] Dauwels J, Srinivasan K, Reddy MR, Musha T, Vialatte F-B, Latchoumane C, Jeong J, Cichocki A. Slowing and loss of complexity in Alzheimer's EEG: two sides of the same coin? Int. J. Alzheimer's Dis. 2011;1–12. doi:10.4061/2011/539621.

[8] Kang Y, Escudero J, Shin D. Principal dynamic mode analysis of EEG data for assisting the diagnosis of Alzheimer's disease. IEEE J. Transl. Eng. Health Med. 2015;3:1–10.

[9] Ghorbanian P, Devilbiss DM, Verma A, et al. . Ann. Biomed. Eng. 2013;41:1243.

[10] Kulkarni NN, Bairagi VK. Extracting salient features for EEG based diagnosis of Alzheimer disease using support vector machine classifier. IETE J. Res. 2017;63(1):11–22.

[11] Staudinger T, Polikar R. Analysis of complexity based EEG features for diagnosis of Alzheimer disease, in: Proceedings of the International Conference on IEEEEMBC, Boston, MA, 2011, pp. 2033–2036.

CHAPTER FOUR

Use of Complexity Features for Diagnosis of Alzheimer Disease

4.1 INTRODUCTION

In previous chapters, we focused on Alzheimer's disease and its progression along with the fundamentals of Electroencephalogram (EEG) signals. We also focused on recording of EEG signal and its characteristics along with significance of each EEG band signal. In this chapter, we will be concentrating on complexity-based features and its analysis on EEG signals for diagnosing patients with Alzheimer's disease. The neuropathology in Alzheimer's disease is mainly concerned with loss of neuronal loss, neurofibrillary tangles, and senile plaques in different widespread regions of brain. These changes basically lead to cognitive as well as behavioral disturbance. The solution to how these neurobiological changes gets converted into functional disturbances can be better understood in terms of a neocortical disconnection syndrome, which results from local neuronal death and deficiency of neurotransmitter. The solution of cortical disconcertion in patients with Alzheimer's disease can be supported by a large number of electrophysiological studies, which demonstrates a tremendous loss of functional interactions using various linear and nonlinear measures [1]. This cortical disconnection not only gives rise to increased complexity in electrophysiological signals [2], but also complexity loss is observed in EEG signals of Alzheimer's disease patients.

Different complexity measures have been used extensively to quantify EEG complexity of mild cognitive impairment (MCI) and patients with Alzheimer's disease. Different complexity measures such as Information theory measures, including Tsallis entropy, approximate entropy, multiscale entropy, auto-mutual entropy, sample entropy [3], multiscale entropy [4], auto-mutual information, and Lempel-Ziv complexity [5,6], and many more are used for analysis of EEG signals considering Alzheimer's disease. Considering physics and information theory, entropy is considered as central quantity. But, entropy is a measure of uncertainty associated with a random variable; this concept was formulated by Shannon, and it is referred to as Shannon entropy. Tsallis entropy is another nonlinear measure for quantifying EEG data by analyzing the variance of the signal in both a slow and rapid manner fashion. Zhao determined the Tsallis entropy by quantizing the amplitude of EEG signal [7]. Approximate entropy highlights that the likelihood patterns in a given signal will not be followed by additional "similar" patterns. Refinement of approximate entropy is the sample entropy, which is designed to have smaller bias. Auto-mutual information of a time series signal represents the mutual information of time series and its time shifted version. The time series is considered to be more complex, if auto-mutual information decays with the corresponding time-lag in faster rate. Lempel-Ziv complexity generally measures the different number of patterns in the signal; the lesser the patterns, in a better way the signals are compressed. In preceding chapter, we will discuss the Lempel-Ziv complexity algorithm and many more in more detail.

In previous literature, complexity measures have also been examined by several physical studies of EEG signal. Fractal dimension [8], correlation dimension, and largest Lyapunov exponent [9] are some of these features that are used for analysis of EEG signal for Alzheimer's disease diagnosis. Fractal dimension is a statistical quantity, which indicates a manner in which a fractal appears completely for filling space as one may zoom down to finer scales. Lyapunov exponent of a system is a physical quantity, which characterizes separation rate of close trajectories. These exponents are related to exponent of dynamical system. This maximal Lyapunov exponent is used for determining the predictability of dynamic system. The positive component indicates that the system is chaotic.

Nonlinear dynamic analysis (NDA) [1] is widely used to various biomedical data to understand complex dynamics of pathological evidence. EEG, on other hand, is the one of the complex biological signal, which requires use of new mathematical tools for better analysis. The application of these NDA algorithms to EEG signals offers significant information considering cortical dynamics. The basic assumption of this NDA is that EEG signals are basically generated using deterministic processes, which are nonlinear and these are coupled together between neuronal populations. Chaotic theory [10] is one of the special techniques that is also used for Alzheimer's disease diagnosis. These systems are irregular and very complex, identical to the stochastic systems. Human Brain is a complex dynamical system, which consists of a large number of interrelated variables, which are difficult to measure directly. Hence, major problem in these systems is measuring and analyzing multidimensional dynamics by knowing only a few variables that is possible to measure. The pathological evidence of decreased EEG complexity on EEG signals is not clear till present. A number of medical studies are still in research to understand the reason for decreased EEG complexity in patients with Alzheimer's disease. But, this decrease in EEG complexity in case of patients with Alzheimer's disease might arise due to neuronal death, deficiency in neurotransmitters such as acetylcholine and connectivity loss of local neuronal networks. Along with factors, dynamical loss of brain responsivity to stimuli can also be responsible for reduced EEG complexity in patients with Alzheimer's disease [11]. Let us discuss the various complexity measures that can be used for determining the EEG complexities in patients with Alzheimer's disease.

4.1.1 Sample entropy

The negative natural log of the conditional probability to the time series of length N, which repeats itself for m data points within a tolerance of r, also repeats for $m + 1$ points.

This entropy is termed as sample entropy [12]. For computation of sample entropy, we can define N^m as the number of matches of length m and N^m_{m+1} as its subset, which matches for length $m + 1$. Sample entropy can be defined mathematically by the following equation:

$$Sample\ Entropy,\ S_{samp} = -\left(\frac{N^m_{m+1}}{N^m}\right) \tag{4.1}$$

Generally the value of m is taken as 2 and r as 0.2 times the standard deviation of the signal. Sample entropy (SE) is elaborated from Approximate entropy, which basically measures the repetition of times series to some extent. Therefore, SE serves to provide an indication of the regularity level or entropy of a time series without making any assumptions. Approximate entropy is already used for diagnosing the Alzheimer's disease, but in recent studies, SE is also used for Alzheimer's disease diagnosis. The previous results indicate that SE is a useful feature that discriminates the AD and moral subjects, and it provides some degree of confidence. This feature was also used in [12] although the approach used was very different. Hence, SE represented the promising feature for predicting the Alzheimer's disease.

4.1.2 Approximate entropy

Approximate entropy is a statistical concept, which is used for quantifying regularity in the data without any prior knowledge about the system [13]. This concept was described by Pincus. Approximate entropy is scale–invariant and it depends on the model to be executed. It can also be applied to short time series data. It is finite for stochastic, noisy, and deterministic processes.

Approximate entropy (ApEn) basically assigns a positive number to time series signal, with large values considering the irregularity in the data. For this purpose, two input parameters must be specified. These parameters are a run length (m) and a tolerance window r. The procedure for obtaining this ApEn is discussed briefly in [14]. This complexity parameter measures the logarithmic likelihood patterns for m contiguous observations, which remain close on further incremental comparisons. ApEn is also used for characterizing different biomedical signals. For instance, it discriminates atypical EEG signals from various counterparts. The use of ApEn for predicting epileptic seizure is discussed in [13,14], but this needs to be clinically validated. Furthermore, ApEn has also been used to quantify the depth of anesthesia [15].

4.1.3 Multiscale multivariate sample entropy

Sample entropy is basically defined as the negative natural log of the conditional probability to the time series signal of length N. In case of multivariate method, user can define the negative natural logarithm of the conditional probability for two composite delay vectors, which are close to each other in dimensional space of p will also be close with each other in space $p + 1$ [6]. The following algorithm gives an idea for computation of this complexity measure: For the time series having p-variate $\{x_{k,i}\}, i = 1, ... n$ as the data series elements, and $k = 1, ... p$ as the number of channels,

   i. Create delay vectors, defined by $X_m(i) = [x_i, x_{i+1}, x_{i+m-1}]$ for $i = 1, ...N-n$, and $n = \max\{M\} \times \max\{\tau\}$, where M represents the embedding vectors and τ the time lag vector.

  ii. Compute the distance of any of the two composite vectors considering the maximum norm.

 iii. Considering standard deviation of the multivariate signal, and fixed threshold r, compute the number of segments P_i, and frequency of occurrence P_i.

 iv. Dimensionality of the vector can be extended from m to $(m + 1)$ by repeating the above steps.

MSMSE can be expressed as,

$$MSMSE(M, \tau, r, n) = -\ln\frac{B^{m+1}(r)}{B^m} \tag{4.2}$$

In the above equation, B^m represents probability for which two composite vectors are in similar in m dimension and $B^{m+1}(r)$ represents probability for which two composite delays will be similar in $(m + 1)$ dimension.

4.1.4 Permutation entropy

Permutation entropy (PE) is one of the fastest as well as robust techniques for extracting information from a time series signal, pertaining to the complexity of the signal [16]. With this feature technique, the time series signal is analyzed for revealing the abnormality of the patient. PE relieves on counting of ordinal patterns in the signal (termed as "motifs"), which describes ups and downs in the signal. The PE is dependent on the measurement of the relative frequencies of various motifs. PE is considered to be the invariant measure; it is expected to quantify system complexity by generating time series and knowing the relative change in the complexity of the data. This entropy is derived from the concept of Shannon entropy to the ordinal pattern analysis with the estimation of relative frequencies, which are extracted from the ordinal patterns of the time series signal. Therefore, it represents the alternative method for measuring the similarities along the patterns with respect to the other such complexity features such as approximate entropy, SE, and many more. In [16], it is highlighted that there exists a high proportion of similar ordinal patterns in the regular time series signal. But, the presence of various different patterns, which occurs similar to the relative frequency, indicates the high complexity level. This permutation entropy is dependent on two parameters: first is the embedding dimension, d and second is the time-lag, τ.

Basically, for a given time series, $y = \{y_1,....y_i,......y_N\}$, it forms data segments in which d represents the number of samples belonging to the segment and τ represents the distance between the sample points that are spanned by the motif section. For computing the permutation entropy of the time series y, series of vectors of length d, $v_{d(n)} = [y_n, y_{n+1},..... y_{(n+d-1)}]$ is computed from signal samples y_j. After computing $v_{d(n)}$, they are arranged in increasing order of magnitudes given by $[y_{n+j1-1}, y_{n+j2-1}, ... y_{n+jn-1}]$. There will be $d!$ possible ordinal patterns, π, for different d samples, which are called "motifs." For each motif π_j, let us assume f(π_j) its frequency of occurrence for given time series. The relative frequency is then given by:

$$p\left(\pi_j\right) = \frac{f\left(\pi_j\right)}{N - d + 1} \tag{4.3}$$

For the fixed dimension $d > 2$, and time-lag $\tau = \tilde{\tau}$, permutation entropy (PE) is given by,

$$H\left(d, \tilde{\tau}\right) = \sum_{\pi_j=1}^{d!} p\left(\pi_j\right) \log_2 p\left(\pi_j\right) \tag{4.4}$$

In above equation, the sum runs over all $d!$ motifs π. The maximum value of $H(d)$ is given by $\log_2(d!)$, implying all motifs have equal probability. The least value of $H(d)$ is zero, which means that time-series signal is very regular.

For the sake of convenience, $H\left(d, \tilde{\tau}\right)$ is normalized by its maximum value $\log_2\left(d!\right)$

$$0 \leq \frac{H\left(d, \tilde{\tau}\right)}{\log_2\left(d!\right)} \leq 1 \tag{4.5}$$

For various selections of the time-lag, τ, computed permutation entropy shows variations in frequency. Therefore, this parameter can be used for analyzing EEG signals for diagnosing Alzheimer's disease, which reflects the slowing effect in EEG signals of patients with Alzheimer's disease.

4.1.5 Multiscale multivariate permutation entropy (MSMPE)

The above complexity measure is suitable for single channel EEG signals; but for multiple channel signals, it is necessary to consider each time series signal separately. But, when it is required to consider the scalp EEG signal, where it is necessary to consider interchannel signal, which can improve the diagnostic performance, it is necessary to use this complexity feature, which is termed as multiscale multivariate permutation entropy [17].

For calculating $MSMPE$, consider the time series window having size T whose sampling frequency is given by $fs = 1/T$. Therefore, each window includes fsT samples, which are termed as data points. For each channel $i \in [1,m]$ and for each $j \in [1, n = d!]$

(i.e., for each motif), compute and count all times $s \in \left[1, \, fsT - d\right]$ for which the channel-time pair (i,s) provides the motif j. The relative frequencies $p_{i,j}$ obtained after dividing the counts by mT, are the entries of the matrix $P_t\,(m,n) = \{p_i, p_j\}$, which reflects the distribution of the motifs in the time series having length T.

As per the above procedure, the multivariate time series signal is converted into the time dependent matrix from which relevant statistics and entropies can be extracted. From the above equations and parameters used, it is easy to calculate marginal relative frequencies, which describes the motif distribution given by,

$$P_j = \sum_{i=1}^{m} p_{i,j} \quad for \quad j = 1,\dots\dots,d! \tag{4.6}$$

The MSMPE is calculated for p_j, for different EEG channels having different complexity are given by:

$$H_{MSMPE}(s) = -\sum_{j=1}^{d!} p_j \log_2 p_j \tag{4.7}$$

From the above matrix, we can also compute the single channel permutation entropy, which is defined by the following equation:

$$H_E(i,s) = -\sum_{j=1}^{d!} mp_{i,j} \log_2\left(mp_{i,j}\right) \quad for \quad i = 1, \dots,m \tag{4.8}$$

This complexity measure can be used for processing of EEG signal to compute this feature on distinct channels. It means that this feature can be computed on different location of brain region to extract the interchannel regularities, which highlights long-range nonlinearities.

4.1.6 Auto-mutual information

MI is the term, which is derived from Shannon's information theory for estimating gained information from various observations of one random event on to another. It measures both linear and nonlinear dependences between two time series signals. Therefore, it is called as nonlinear correlation function. Hence, this parameter can also be applied to the time-delayed versions having same sequences. The mathematical function used to calculate auto-mutual information is discussed in [18]. It is discussed in [18] that rate of auto-mutual information decreases with increasing delay in time, which correlates with the signal entropy. The decreased auto-mutual rate is correlated with signal entropies for increasing time delays. This concept is used for characterizing EEG signals for comparing subjects with Alzheimer's disease with normal subjects. This parameter was also used to study schizophrenia.

4.1.7 Tsallis entropy

The variance of signal analyzed by both slow and rapid manner in nonlinear method for quantifying the EEG signal is termed as "Tsallis Entropy" [19]. Eqs. (4.9) and (4.10) are computed by using MATLAB software and represents slow and rapid variances for EEG signal. The slow variance is only the measurement of variance throughout the entire signal (or either the epoch of signal being analyzed), and the rapid variance represents the variance from each critical point to the next. Critical points direct to local maxima and minima in the data. Tsallis entropy value (qEEG), on the other hand, is simply the ratio of all rapid variances divided by the slow variance and then subtracted from one as shown in Eq. (4.11). Tsallis entropy has been shown to be a good analysis method to use with working memory tasks [19].

$$var_{slow} = \sum_{i=1}^{N} (x_i - \overline{x})^2 \tag{4.9}$$

$$var_{Rapid} = \sum_{\substack{j=1 \\ x \in Interval\ j}} (x_i - \overline{x})^2 \tag{4.10}$$

$$qEEG = 1 - \frac{\sum_{Interval\ j} var_{Rapid}}{var_{slow}} \tag{4.11}$$

4.1.8 Lempel-Ziv complexity

This complexity measure was suggested by Lempel and Ziv, for sequences of finite length. It is simple for calculation and is a nonparametric measure of complexity in case of one-dimensional signal, which does not require long data sample for computation [5]. This complexity measure is related to the number of substrings and its recurrence rate for the given sequence, with larger values, which corresponds to the more complexity in the data. This complexity feature is used for studying brain function, transmission of brain information, and many more. It is also used for quantifying depth of anesthesia.

Lempel – Ziv complexity is based on a coarse-graining of measurements. Therefore, prior to the computation of complexity measure c(n), the signal needs to be transformed into a finite symbol sequence. Let us study the following two sequence conversion methods [15]:

a. **0–1 sequence conversion**: In this sequence conversion, median value is considered as the threshold T_d. Comparing with the threshold, the time domain signal is converted into a 0–1 sequence $P = s(1), s(2),....,s(n)$ with $s(i)$ given by following equation:

$$s(i) = \begin{cases} 0 \ if \ x_i < T_d \\ 1 \ if \ x_i \geq T_d \end{cases} \tag{4.12}$$

b. 0–1–2 sequence conversion: In this conversion mode for each segment of EEG signal, median value λ_m, maximum value λ_{max} and minimum value λ_{min} are computed. In this case, two thresholds are obtained: $T_{d1} = \lambda_m - |\lambda_{min}|/16$ and $T_{d2} = \lambda_m + |\lambda_{max}|/16$ []. After computing values, EEG data are then converted into 0–1–2 sequence, $P = s(1), s(2), \ldots, s(n)$, with $s(i)$ defined by following equation:

$$s(i) = \begin{cases} 0 \ if \ x_i \leq T_{d1} \\ 1 \ if \ T_{d2} < x_i < T_{d2} \\ 2 \ if \ x_i \geq T_{d2} \end{cases} \qquad (4.13)$$

The P sequence is scanned in both the conversion methods from left to right and the complexity counter $c(n)$ gets increased by one unit in every time sequence as a new subsequence of consecutive characters is encountered.

For computing the complexity measure, which does not depend on the sequence length, value of $c(n)$ needs to be normalized. Considering n as the length of the sequence and α as the number of different symbols, one might consider value of $\alpha = 2$. The upper bound of $c(n)$ is given by the following expression:

$$b(n) = \frac{n}{(1 - \varepsilon_n)\log_\alpha n} \qquad (4.14)$$

In above equation, ε_n is small quantity, that is, $\varepsilon_n \to 0$, when the value of n is large $n \to \infty$, then the above formula becomes,

$$b(n) = \frac{n}{\log_2 n} \qquad (4.15)$$

Therefore, the value of $c(n)$ can be normalized by $b(n)$, to get normalized Lempel – Ziv complexity $C(n)$, which does not depend on the value of sequence:

$$C(n) = \frac{c(n)}{b(n)} \qquad (4.16)$$

In [17], Multiscale multivariate Lempel Ziv complexity (MSMLZ) results were computed by averaging complexity values for all possible electrodes analyzed belonging to a cerebral area for each scale factor. But, in some studies, normalized LZ complexity was computed for the sequence of EEG data, which captures the temporal structure of the sequence.

4.2 USE OF NEW COMPLEXITY FEATURES IN ALZHEIMER'S DISEASE DIAGNOSIS

Section 4.1 summarizes the previously used several nonlinear dynamical as well as complexities measures used for analyzing of EEG signals for diagnosis of various neurological disorders such as Alzheimer's disease, epilepsy, brain stroke, and many more. But, in this section, we focus on new complexity measures that were used for investigation. Let us study the new proposed complexity features in detail.

4.2.1 Spectral entropy

Spectral entropy indicates the amount of unpredictability and disorder in spectrum of EEG. Higher complexity is achieved if higher amount of spectral is entropy is observed [9,10]. It is computed in the following manner:

1. For the given signal $x(t)$, compute $S(f)$, the power spectral density (PSD), as the Fourier transform of the autocorrelation function of the signal $x(t)$.
2. Depending upon the frequency of interest, extract the power in the spectral band from 0.5 to 30 Hz.
3. After calculation of the spectral band power, normalize the power in the given band of interest.
4. Calculate the spectral entropy by using formula,

$$SE = \sum_{f=0,5}^{40} S(f) \star \ln \frac{1}{s(f)} \tag{4.17}$$

5. Compute the spectral entropy as given above.

4.2.2 Spectral centroid

Spectral centroid measures the shape of the spectrum of EEG signals. A higher value of SC corresponds to more energy of the signal being concentrated within higher frequencies. Basically, it measures the spectral shape and position of the spectrum [20].
It is computed as follows:

1. Let $x_i(n)$, $n = 0,1,...N-1$ be the sample of the ith frame, with $X_i(k)$, $k = 0,1,...N-1$, as the Discrete Fourier Transform (DFT) coefficients of the sequence.
2. Compute the Spectral centroid of the each frame as:

$$C(i) = \frac{\sum_{k=0}^{N-1} k |Xi(k)|}{\sum_{k=0}^{N-1} |Xi(k)|} \tag{4.18}$$

The mean value of the spectral centroid across all the frames can be used as the SC feature for each epoch of the frame of the EEG Signal.

4.2.3 Spectral roll off

Spectral roll off represents the frequency below which a certain percentage (usually 80%–90%) of the magnitude distribution of the spectrum is concentrated in the spectrum. It is computed as follows:

1. Let $x_i(n)$, $n = 0,1,...N-1$ be the sample of the ith frame, with $X_i(k)$, $k = 0,1,...N-1$, as the Discrete Fourier transform (DFT) coefficients of the sequence.

2. Compute the spectral roll off as the sample that satisfies,

$$\sum_{k=0}^{k(i)}|X_i(k)| = \frac{P}{100}\sum_{k=0}^{N-1}|X_i(k)| \tag{4.19}$$

where the **P** parameter is sometimes chosen between 80 and 100.

4.2.4 Zero crossing rate

Zero crossing rate of any signal frame is the rate at which a signal changes its sign during the frame. It denotes the number of times the signal changes value, from positive to negative and vice versa, divided by the total length of the frame.

Zero crossing rate is computed as:
1. Let $x_i(n) = 0,1,....N-1$ be the samples of the ith frame.
2. Zero crossing rate of each frame is calculated as:

$$Z(i) = \frac{1}{2N}\sum_{n=0}^{N-1}\left|sgn[x_i(n)] - sgn[x_i(n-1)]\right| \tag{4.20}$$

where

$$sgn[x_i(n)] = \begin{cases} 1, & x_i(n) \geq 0 \\ -1, & x_i(n) < 0 \end{cases}$$

4.3 DISCUSSION AND CONCLUSION

The complexity of EEG signal can be quantified extensively by the use of above techniques to reveal several diagnostic information for characterizing several brain electrical activities. In [21], it is highlighted that PE shows the changes in the transition between inter-ictal and ictal states in epileptic brains by using complexity measures. The same complexity can be used for revealing abnormalities such as Alzheimer's disease. In [6], it is also highlighted that larger values of Lempel Ziv complexity signify more complexity of signals, that is, there are more chances of new pattern generation. It is concluded that brains affected by Alzheimer's disease depict less complex electrophysiological behavior in occipital, parietal, and temporal regions. Many studies showed similar findings for EEG signals of patients with Alzheimer's disease [6,14,17]. The implications of these neurophysiological studies of reduced complexity are yet not clinically validated. Some of the reasons for these may be the (a) death of neuronal cells, (b) deficiency in the neurotransmitter, and (c) loss of connectivity in neural networks of the nerve cell in brain regions. The decreased LZ complexity in case of EEG is termed as "decomplexification" as it is associated with inactivation of neural networks. Model independence is another feature of LZ complexity that is discussed in [17].

The complex nature of EEG signals therefore requires new signal processing as well as mathematical methods for investigating cortical dynamics. NDA may provide useful technique for understanding brain activities. Association of nonlinear EEG dynamics and cognitive performance, longitudinal changes in nonlinear dynamics, and functional connectivity among several cortical areas affected by Alzheimer's disease needs to be examined and studied. NDA analysis may also help for improving the diagnostic accuracy and early detection of Alzheimer's in preclinical stages. It is expected that nonlinear dynamic analysis may contribute for deeper understanding of physiological phenomenon in case of Alzheimer's disease, which is not possible by typical spectral analysis.

The multiscale and multivariate entropic complexity measures are also capable of processing EEG data, which are capable of distinguishing between various brain states. These features are also capable to capture slowing effect related to the Alzheimer's disease. Entropy is therefore considered as the powerful technique that is used for analysis of physiological time series signal since it describes distribution of probabilities for complex system such as EEG in present application. The entropic measures are also capable of modeling the dynamical couplings between several variables relating to the underlying cause. Due to this, permutation entropy is suitable for monitoring the changes in brain regions to distinguish between ageing and dementia. Therefore, the potential of using various such entropies can be considered as a complementary approach for diagnosis of Alzheimer disease. In future, other nonlinear dynamic analysis techniques such as Central tendency measure and compressive sensing approach can be explored for analysis of EEG signal for Alzheimer's disease diagnosis (Table 4.1).

In Ref. [20], different complexity measures were analyzed to observe and study them if they carry any diagnostic information for diagnosis of Alzheimer disease. In medical concept, it is signified that AD affects the neuronal activity of the patients. The features

Table 4.1 Studies concerning with reduced complexity and nonlinear dynamical analysis of EEG for Alzheimer's disease diagnosis

Sr. No.	Name of the complexity measure	Reference
1	Sample entropy	[3,4]
2	Approximate entropy	[3,13,14]
3	Multivariate multiscale sample entropy	[3,6,18]
4	Permutation entropy	[6,16]
5	Multivariate multiscale permutation entropy	[6,16,17]
6	Auto-mutual entropy	[18]
7	Tsallis entropy	[19]
8	Lempel Ziv complexity	[1,3,5,6]
9	Multivariate multiscale Lempel Ziv complexity	[6]
10	New proposed complexity measures (features)	[20,22]

used in Refs. [20,22] show decreased complexity values for patients with AD. The difference in the complexity-based feature values among the cohort is small, but indicates its significance on the electrodes of EEG. The AD group features consist of lower values, suggesting that patients with AD tend to be less complex. The features used carry significant information in the central, parietal, temporal, and frontal lobe. This reduced complexity is observed in patients with Dementia and Alzheimer's disease due to the appearance of the neurofibrillary plaques and tangles in cortex region. Complexity feature values were also lower for patients with Alzheimer's disease in the frontal, central, and temporal lobes. Higher amount of spectral content is seen in higher frequencies for Normal group. This is predicted as the high level of complexity in Normal subjects. It is to highlight that when we combine these features with one another they can provide more diagnostic information and thus high classification rates are obtained.

The reason for slowing of EEG signals in case of patients with Alzheimer's disease may be due to the increase in the neuronal connectivity in the brain cells and degeneration of beta amyloid protein. The level of neprilysin gets increased due to the stem cell. It is the enzyme which breaks the level of amyloid proteins and lowers the brain activity in case of patients with Alzheimer's disease. Due to this, an alpha and beta activity also reduces and complexity gets reduced since signal becomes slow [20,22].

SUMMARY

In this chapter, we emphasized on different complexity features as well as the several nonlinear features that are used for EEG signal analysis for detection of Alzheimer disease. It is highlighted that EEG signals of patients with Alzheimer's disease slow down as compared to the healthy patients. The feature carries significant information in the central, parietal, as well as frontal brain regions. This reduced complexity is observed due to the presence of neurofibrillary plaques and tangles in the cortex regions, neuronal loss of cells and death of neurons. Information theory and nonlinear dynamic analysis play significant role in EEG analysis. Various different entropy measures were discussed in the chapter highlighting its use in Alzheimer's diagnosis. These features may be further investigated to obtain better results in terms of classification as well as diagnostic accuracy along with understanding the brain dynamics.

REFERENCES

[1] Jeong J. EEG dynamics in patients with Alzheimer's disease. Clin Neurophysiol 2004;115:1490–505.
[2] Stam CJ, van der Made Y, Pijnenburg YA, Scheltens P. EEG synchronization in mild cognitive impairment and Alzheimer's disease. Acta Neurol Scand 2003;108:90–6.
[3] Richman F S, Moorman F R. Physiological time-series analysis using approximate entropy and sample entropy. Am J Physiol (Heart Circ Physiol) 2000;274:2039–49.
[4] Costa M, Goldberger AL, Peng C-K. Multiscale entropy analysis of physiologic time series. Phys Rev Lett 2002;89:062102.
[5] Lempel A, Ziv J. On the complexity of finite sequences. IEEE Trans Inf Theory 1976;22:75–81.

[6] Labate D, La Foresta F, Morabito G, Palamara I, Morabito FC. Entropic measurement of EEG complexity in Alzheimer's disease through a multivariate mulstiscale approach. IEEE Sens J 2013;13:3284–92.

[7] Zhao P, Van-Eetvelt P, Goh C, Hudson N, Wimalaratna S, Ifeachor E. Characterization of EEG's in Alzheimer's disease using information theoretic methods. In: Mandelbrot BB, editor. Conference Proceedings of 2007 IEEE Engineering in Medicine and Biology Society. W.H. Freeman and Company; 1982. The Fractal Geometry of Nature.

[8] Mandelbrot BB. The Fractal Geometry of Nature. W.H. Freeman and Company; 1982.

[9] Pesin YB. Characteristic Lyapunov exponents and smooth ergodic theory. Russ Math Surv 1977;32(4):55–114.

[10] Soong ACK, Stuart CIJM. Evidence of chaotic dynamics underlying the human alpha-rhythm electroencephalogram. Biol Cybern 1989;62:55–62.

[11] Pritchard WS, Duke DW, Coburn KL, Altered EEG. Dynamical responsivity associated with normal aging and probable Alzheimer's disease. Dementia 1991;2:102–5.

[12] Lake DE, Richman JS, Griffin MP, Moorman JR. Sample entropy analysis of neonatal heart rate variability. Am J Physiol 2002;283(3):R789–97.

[13] Pincus SM. Approximate entropy as a measure of system complexity. Proc Natl Acad Sci USA 1991;88:2297–301.

[14] Abásolo D, Hornero R, Espino P, Poza J, Sánchez CI, de la Rosa R. Analysis of regularity in the EEG background activity of Alzheimers disease patients with approximate entropy. Clin Neurophysiol 2005;116(8):1826–34.

[15] Zhang XS, Roy RJ, Jensen EW. EEG complexity as a measure of depth of anesthesia for patients. IEEE Trans Biomed Eng 2001;48:1424–33.

[16] Bandt C, Pompe B. Permutation entropy: a natural complexity measure for time series. Phys Rev Lett 2002;88:174102.

[17] Morabito FC, Labate D, La Foresta F, Bramanti A, Morabito G, Palamara I. Multivariate multi-scale permutation entropy for complexity analysis of Alzheimer's disease EEG. Entropy 2012;14(7):1186–202.

[18] Jeong J, Gore J, Peterson B. Mutual information analysis of the EEG in patients with Alzheimer's disease. Clin Neurophysiol 2001;112:827–35.

[19] De Bock T, Das S, Mohsin M. Early detection of Alzheimer's disease using nonlinear analysis of EEG via Tsallis Entropy. Published in IEEE Biomedical Sciences & Engineering Conference May 2010;1–4.

[20] Kulkarni NN, Bairagi VK. Extracting salient features for EEG based diagnosis of Alzheimer disease using support vector machine classifier. IETE J Res 2017;63(1):11–22. Published by Taylor and Francis (UK).

[21] Mammone N, Lay-Ekuakille A, Morabito FC, Massaro A, Casciaro S, Trabacca A. Analysis of absence seizure EEG via permutation entropy spatio-temporal clustering. (MEMEA 2011) Bari, Italy. Proceeding of IEEE International Symposium on Medical Measurements and Applications, 30–31. p. 532–5.

[22] Staudinger T, Polikar R. Analysis of complexity based EEG features for the diagnosis of Alzheimer's disease. Conf Proc IEEE Eng Med Biol Soc 2011;2011:2033–6.

FURTHER READING

[23] Blurton-Jones M, Spencer B, Michael S, Castello NA, Agazaryan AA, Davis JL, Muller FJ, Loring JF, Masliah E, LaFerla FM. Neural stem cells genetically-modified to express neprilysin reduce pathology in Alzheimer transgenic models. Stem Cell Res Ther 2014;5:46.

CHAPTER FIVE

Classification Algorithms in Diagnosis of Alzheimer's Disease

5.1 INTRODUCTION

Artificial intelligence (AI) is the ability of a computer system or actual complex algorithms to estimate relationships of the complex datasets and also to find correlations, which may yield testable hypotheses. Machine Learning, on the other hand, is the ability of the computer systems to incorporate these newly generated data for future assessments [1,2]. The method of programming by humans for all these processes makes the system more efficient. Programming of these Machine Learning algorithms makes Machine Learning more efficient and better. AI in healthcare applications has already spanned the total clinical experience, but it has its used and typical application in the following key areas [1–3]:

- Data Mining: It consists of gathering meaningful information by method of surveying huge quantity of data or analyzing it; hence, it is called Big Data.
- Medical Imaging: In this application, interpretation of medical images is done to sort different patterns otherwise obfuscated by indiscernible noise.
- Clinical Decision Support: In this mode of application, data is gathered from patients, electronic records, and many more. The evaluative testing is sorted into different probabilities, which have some possibilities, and it has some options, which are likely to be correct, considering the patient, which can have optimal benefit.

Recent advancement in Machine Learning has been laid by the development of new learning algorithms, theory, and by the ongoing explosion in the availability of online data and low-cost computation. The maintenance of these data-intensive Machine Learning methods can be explored throughout technology and commerce, which can lead to high evidence-based decision-making across many situations, including various applications such as healthcare applications, manufacturing systems, education, financial modeling, and many more. Machine Learning has made rapid progression over the past two decades apart from transformation of practical technology to commercial use. As a part of AI technology, Machine Learning algorithms have emerged as the tool of choice for developing practical software systems for computer vision applications, speech recognition systems, natural language processing, robotic control systems, and other applications, including biomedical applications [4,5]. It highlights that training a system for desired input–output behavior is easier rather than

EEG-Based Diagnosis of Alzheimer Disease
http://dx.doi.org/10.1016/B978-0-12-815392-5.00005-8

programming it manually by forecasting the desired response for all inputs. Machine Learning methods have emerged and have been developed for analyzing experimental data, which have high throughput.

For Machine Learning algorithms to be advantageous in solving medical diagnostic tasks, different parameters are to be considered, namely good performance of the classifier or proposed classification algorithm, ability to deal with the missing data, as well as noisy data, transparency of the diagnostic knowledge along with ability to explain decisions, and, lastly, ability of the algorithm to reduce the number of tests necessary to obtain reliable diagnosis. Let us discuss these key requirements of the algorithms in a more detailed manner [3].

1. **Good Performance of an algorithm**: The proposed algorithm should be able to extract significant information from available data during classification. The classification or diagnostic accuracy on either new class or cases should be as high as possible. Therefore most of the algorithms should perform better considering the classification accuracy when using the same description of patients. Several Machine Learning techniques should be tested on the available data for classification, and the one or few with the best-estimated performance should be considered for the development of the application.

2. **Dealing with the missing data**: In medical diagnosis systems, it is quite often that description of patients is lacking from certain database. Therefore ML algorithms should be able to deal with such missing data and incomplete records during classification.

3. **Appearance of Noisy Data**: Basically, a lot of errors and uncertainty is present in medical data. Hence, Machine Learning algorithms used for medical applications should have effective techniques for handling noisy data.

4. **Transparency of diagnostic data**: The decision obtained and generated knowledge should be translucent to the physician. The physician should be able to analyze and understand the generated knowledge. In general, the generated knowledge should provide the physician a new direction, and it can also reveal new interrelations between the obtained results.

5. **Explanation ability**: The system should be able to explain the decisions when diagnosing new patients. When the system deals with the unexpected solution, the physician needs further explanations to understand the system's suggestions.

6. **Reduction in the number of the states:** In medical diagnosis, data collection is expensive and time consuming. Hence, it is necessary to have a classifier that can reliably classify a small amount of data. This can be tested by providing each algorithm with limited amount of data. But, the method of determining the appropriate subset of data is time consuming. Some of the Machine Learning algorithms can select an appropriate subset of attributes. It means that selection of parameter is done at the time of learning process, which can reduce the number of the states.

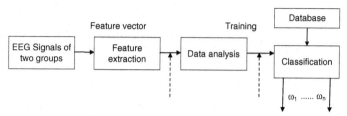

Figure 5.1 *Supervised statistical pattern recognition flow.*

Machine Learning is a technique of programming the computer system to optimize a performance criterion using sample data or past experience. In Machine Learning algorithms, a model is defined by some parameters, and then learning is executed using a computer program to optimize the parameters of the model using trained data or past experience. The model can be predictive to make future predictions, or it can be also descriptive to gain knowledge from data, or both. Statistical theory is used in Machine Learning for building mathematical models, since the core task is making inference from a sample database [5,6]. The performance of computer science is twofold. In training mode, efficient algorithms are required for solving the optimization problem, as well as for storing and processing of extensive amount of data. Secondly, once a model is learned completely, its representation, as well as algorithmic solution for inference needs to be efficient as well. In some of the applications, efficiency of the learning algorithm, predictive accuracy, space, and complexity of the inference algorithm are also important.

Pattern recognition is one of the most important aspects of AI; it is also an appropriate field for the development, data validation, and comparison of different learning techniques: statistical or structural, supervised or unsupervised, and inductive or deductive. In the present study, supervised learning framework is incorporated, which requires a large database for classification purpose [4,6]. This database holds labeled patterns belonging to the X predefined classes $\omega_1, \omega_2 \ldots \omega_X$. These patterns are repeatedly presented to the classifier in order to derive a decision rule optimizing a given criterion. Fig. 5.1 shows the idea of supervised classifier technique.

In the present research work, database consists of two groups consisting of patients with Alzheimer's disease and normal patients; it is required to classify the patients between these two groups. Therefore it is necessary to implement Machine Learning algorithms for data classification in two groups. In the present work, 100 samples of EEG data are used; out of which 50 samples consist of Alzheimer's patients, while the remaining 50 samples consist of normal patients. In the present work, 50% of the data is used for training purpose, while the remaining 50% data is used for testing purpose. In literature, different classifiers are used for data classification, namely Linear Discriminant Analysis (LDA), Neural Networks (NN), Random forests, etc. But in the present study, Support Vector Machine (SVM) & K-Nearest Neighbor (KNN) classifiers are used for classification of data in two groups. It is seen that in case of EEG classification problem, dimensional space is high, since dimensions of the patterns increase with the number of features selected. On the other hand, MATLAB software provides the efficient tools and methods for classification purpose, in which the classification of two data groups can be done efficiently by use of in-built functions. Classifiers are also used in various applications such as data mining, pattern matching, and pattern recognition. Some classifiers are implemented by use of special toolboxes in variety of applications; for example, WEKA source toolbox is available for classification using JAVA platform. In our study, we have implemented the MATLAB-based classification for distinguishing between two groups. Let us study the classifiers in detail.

5.1.1 Support Vector Machine

SVM is a supervised Machine Learning algorithm, which is largely used for classification and regression purpose. SVM conceptually implements the following logic: input vectors are either linearly or nonlinearly mapped into a high-dimensional feature space. Linear decision surface is constructed in this dimensional feature space. High generalization ability of the Machine Learning is ensured in linear decision surface [10]. In simple words, it can be explained as, given a set of training examples, each sample is marked with the category that it belongs to. SVM training algorithm builds up a logic model, which predicts whether a new example falls into one category or into the other category. SVM model is a representation of the examples as a point in the feature space. Examples are used for the find out the gap between data to split them to the different region for different classes. Test data are then mapped into this space and predicted to belong to a category based on which side of the gap they fall on [7,8].

The technical problem of implementing this algorithm is to computationally treat such high-dimensional spaces; for instance, to construct polynomial of degree 4 or 5 in a 200-dimensional space, it can be important to construct hyperplanes in a million-dimensional feature space. The aforementioned conceptual was resolved in 1965 [8] for the case of optimal hyperplanes for separable classes. An optimal hyperplane is defined as the linear decision function, which has maximal margin between the vectors of the two classes. It was seen that for constructing such optimal hyperplanes for classification purpose, it is necessary to take into account a small amount of the training data, which are called support vectors, which determine the margin.

5.1.1.1 Linear separation

In linear separation mode of classification using SVM, firstly, L training point, where each input x_i has D attributes and is in one of the classes,

$$\{x_i, y_i\}, \; i = 1....L, \; y_i \in \{1, -1\}, \; x \in R^D \tag{5.1}$$

The aforementioned training dataset is assumed linearly separable, which means that straight line on a graph separated the two or more points in two different classes. This

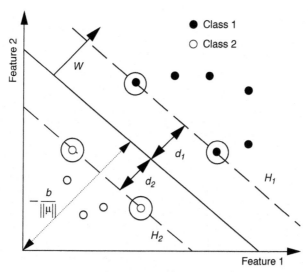

Figure 5.2 Hyperplane of two linearly separable classes [9].

hyperplane can be explained by the equation $w \cdot x + b = 0$. In this linear system, w stands for normal vector to the hyperplane, and the perpendicular distance from the hyperplane to the origin point is given by $\dfrac{b}{\|w\|}$. Support vectors are the examples closest to the separating hyperplane, and the objective of the SVM classifiers is to orientate this hyperplane in such a way to be as far as possible from the closest members of the classes as shown in Fig. 5.2 and 5.3.

It can be observed in Fig. 5.2 that the points are separated within the line crossing between them. Each point provides one of the equations below, so that it can be determined for which class it belongs to.

$$x_i \cdot w + b \geq +1 y_i = +1 \tag{5.2}$$

$$x_i \cdot w + b \leq -1 y_i = -1 \tag{5.3}$$

Lines passing between data limit the interval of classes. Distance between lines H_1 and H_2 can be also higher for better classification. This length of SVM margin can be maximized. By using concept of vector geometry, it is proved that the margin is equal to $\dfrac{1}{\|w\|}$, and maximizing it can also be similar to minimizing $\|w\|$. For minimum, value of $\|w\|$ equation is also true:

$$y_i (x_i, w + b) - 1 \geq 0 \quad for \ \forall_i \tag{5.4}$$

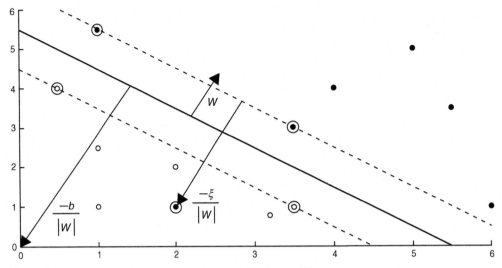

Figure 5.3 *Hyperplane of two nonlinearly separable classes [9].*

5.1.1.2 Nonseparable classes

In some cases, data might not be allowed for a separating hyperplane. A concept of soft margin is used, in which hyperplane separates many but not all data points. In general term, there are two standard formulations for soft margin. It basically consists of slack variable s_i and penalty parameter C. The L^1-norm problem is described by the following expression:

$$\min_{w,b,s} \left[\frac{1}{2} \, w,w + C \sum_i s_i \right] \qquad (5.5)$$

such that $y_i \left(w_i, x_i + b \right) \geq 1 - s_i,\ s_i \geq 0$.

The L-norm refers to using s_i as slack variables instead of their squares (Fig. 5.3).

The L^2-norm problem is given by:

$$\min_{w,b,s} \left[\frac{1}{2} w,w + C \sum_i s_i^2 \right] \qquad (5.6)$$

subject to same constraints.

5.1.1.3 Nonlinear transformation with kernels

Some binary classifiers may not have a simple hyperplane as a useful separating principle. For this kind of problems, a mathematical approach is formulated, which in turn returns nearly all the simplicity of support vector separating hyperplane. This approach is generally used for obtaining precised results from the method of reproducing kernels.

There exists a class of functions $k < x,y>$ with the following problem, which consists of linear space S and a function mapping x to S, such that

$$k(x,y) = \langle \varnothing(x), \varnothing(y) \rangle$$

The dot product takes place in the same space S. This class of function includes the following:

1. Polynomials: For some positive integer d, $k(x,y) = \left(1 + \langle x,y \rangle\right)^d$.
2. Radial basis function: For some positive number σ, $k(x,y) = \exp(- < (x-y),(x-y) \geq \left(2\sigma^2\right)$.
3. Multilayer Perceptron (Neural Network): For a positive number p_1 and a negative number p_2, $k(x,y) = \tanh\left(p_1 \langle x,y \rangle + p_2\right)$.

The mathematical approach depends on computational method of hyperplane by use of kernel function. Dot product technique is specifically used for obtaining calculations for classification purpose using hyperplane technique. Therefore a nonlinear kernel makes use of identical calculations, algorithmic solutions & classifiers are obtained which are nonlinear in nature. The obtained classifiers are hypersurfaces in some space S (say), but the space S may not be examined or identified. In the proposed research work, supervised learning model is used. Firstly, a Support Vector Machine is trained, and then uses the trained machine to classify new data. In order to obtain satisfactory or rather better classification accuracy, a variety of SVM kernel functions are used. In programming approach, the SVM classifier is trained using "svmtrain" function. The most common function used is the following:

SVMstruct = svmtrain (data, group, "kernel function," "rbf").

The inputs for the aforementioned function are the following:

- Data: Matrix of data points, where each row represents one observation, and each column represents one feature.
- Groups: Groups represent the column vector with each row corresponding to the value of the corresponding row in data group can have logical entries, or can be a double vector or cell array with two values.
- Kernel function: The default value of "linear" separates the data by a hyperplane. The value "rbf" uses a Gaussian radial basis function. Hence, SVMstruct function contains the optimized parameters from the SVM algorithms, enabling classification of new data.

5.1.1.4 Advantages of Support Vector Machine in classification of EEG signal

Real-time system classification problems mainly deal with the difficulty in high-dimensional spaces, various numbers of classes and time, as well as memory management problems. SVM classifier generally works on high-dimensional spaces. In data classification of EEG signals for different applications, feature dimensional space is very high, since dimensions of the patterns increase with the number of features selected. It also depends on the sensor channel used for acquisition. Another point which is to be noted here is, SVM classifier generally uses overfitting protection, which generally may not depend on the number of features, and it has the potential to handle or rather overcome these large feature spaces for purpose of data classification.

For classifying data consisting of many classes, SVM classifier is generally used. In such cases, different labels are used for representing each class of data. Therefore classification problem can be avoided for data classification having different classes. The dominant approach is incorporated for reducing the single multiclass problem to a number of multiple binary classification problems. Herein, each of the problems results in a binary classifier, which can be assumed to result in a output function that relatively gives large values for obtaining examples from positive class and small values for examples that belong to negative class.

5.1.2 K-Nearest Neighbor classifier (K-NN classifier)

KNN classifier is a semi-supervised classifier used for classification of data in two or more groups. The classification of data using this classifier is performed by comparing the new sample data termed as testing data with the baseline data, which is referred to as a training data. It searches the neighborhood element K in the training data, and then the class is assigned, which appears more periodically in the neighborhood of K. In this classification approach, it is very important to vary the value of K in order to obtain the match class between training and testing data. In general, default value of K is considered or rather taken to be 1. But, for obtaining more classification accuracy, this K value can be varied from K = 1, 2, 3, up to 10. The neighborhood setting used for distance measurement is set to "Euclidean" or "nearest". This Euclidean distance metric is used for finding the similarity of object in the neighborhood of K [9,10]. To calculate the Euclidean distance, the following formula is used:

$$d\left(X_i, Y_j\right) = \sqrt{\sum_i \left(X_i - Y_j\right)^2} \tag{5.7}$$

In general, KNN algorithm is one of the Machine Learning algorithms with the following features:
- A technique for classification of data or objects dependent on closest training examples in the feature space.
- The function is approximated locally. Due to this, the algorithm is sensitive to the local structure of the data.
- All data computations are deferred until classification.

In this Machine Learning algorithm, data with input x can be classified by using majority vote of its neighbor. It is specifically assigned to most common class along K-Nearest Neighbors. By default, value of K is equal to 1; then data is simply assigned to the nearest neighbor class. This is referred to as "$1-n$" classification. The drawback of this basic majority voting is that the classes having more frequent examples dominate the prediction, since they come up with nearest neighbors. The only way to overcome such a problem is to scale the classification by taking into account the distance between the test point to each of its L-Nearest Neighbors.

5.1.2.1 Estimating K-Nearest Neighbor value

The following relation is used for estimating K-Nearest Neighbor value,

$$(x) = \frac{1}{k} \sum_{x_i \in N_k(x)} y_i \in A_{VE}\{y_i \mid x_i \in N_k(x)\} \tag{5.8}$$

where $N_k(x)$ consists of the k closest points to x in the sample.

Considering the binary classification, it is estimated as:

$$\gamma(x) = \frac{1}{k} \sum_{x_i \in N_k(x)} y_i \in [0,1]$$

$$f(x) = \begin{cases} 1 \ if \ \gamma(x) > 0.5 \\ 0 \ if \ \gamma(x) < 0.5 \end{cases}$$

In binary (two-class) classification problems, it is important to choose value of "K" as an odd number in order to avoid tied votes. The advantages of using K-Nearest Neighbor classifier in biomedical signal processing applications includes the following:
1. stringent assumptions are not relied considering the data,
2. low bias,
3. robust against outliers, and
4. 1-nn can be very useful in low-dimensional problems.

In the present research work, "knnclassify" MATLAB function is used for classification of data between two groups using K-NN classifier.

Class = Knnclassify (Sample, Training, Group)

The aforementioned function classifies each data in a row in sample into one of the groups in training mode using the nearest neighbor method. Sample and training are the matrices with the same number of columns are used for comparison. Group is a grouping variable used during training. The unique values define the groups, and each element is then defined by the specific group to which the corresponding row of training belongs. This group value element can be a numeric vector, an array of string, or a cell array of strings. It is necessary to have training and group with the same number of rows. The classifier treats NaNs (Not a number) or empty strings in a group, since it consists of missing values, and therefore it ignores the corresponding rows of training. The "class" variable indicates the assigning of sample data to each group or rather whether it is of same type of group. Let us see the class variable syntax used in MATLAB.

1. Class = Knnclassify (Sample, Training, Group, K) allows the user to specify K value, the nearest neighbor value used in the classification purpose. The default of K is 1.
2. Class = Knnclassify (Sample, Training, Group, K, Distance) allows the user to select the distance metric.

The following choices are made during classification:

1. "euclidean"—Euclidean distance (default).
2. "cityblock"—Sum of absolute differences, or L1.
3. "cosine"—One minus the cosine of the included angle between points (treated as vectors).
4. "correlation"—One minus the sample correlation between points (treated as sequences of values).
   • Class = Knnclassify (Sample, Training, Group, K, Distance, Rule) allows the user to specify the rule used to decide how to classify the sample.

The following choices are made during classification:

1. "nearest"—Majority rule with nearest point tiebreak.
2. "random"—Majority rule with random point tiebreak.
3. "consensus"—Consensus rule.

In data classification using KNN classifier, majority rule is used as a default. In this, a sample point is assigned to the class for which K–Nearest Neighbor elements are available. "Consensus" is typically used for acquiring consensus, as opposition to majority rule. When we are using consensus option, the points that are not of all the same class are not assigned. When it is required to classify more than two groups, it is necessary to break the tie in the nearest neighbors. For breaking this tie, "random" and "nearest" are the two options that can be used for breaking this tie. Nearest tiebreak is the default majority rule used commonly. In the next chapter, use of KNN classifier in data classification is explored briefly.

SUMMARY

The development of Machine Learning algorithms and its applications in medical diagnosis have emerged leading to the advancement in clinical data analysis. In future technology, intelligent data analysis can be used due to the production of huge amount of data. Currently available Machine Learning algorithms help medical researchers and practitioners for revealing significant relationships in their data analysis. We also focused on Support Vector Machine and K–Nearest Neighbor classifier, which are used for data classification in the next chapter, in a detailed manner.

REFERENCES

[1] Jordan MI, Mitchell TM. Machine learning: trends, perspectives, and prospects. Science 2015;349:255.
[2] Shavlik JW, Dieterich TG, editors. Readings in machine learning. Morgan Kaufmann; 1990.
[3] Kononenko I. Machine learning for medical diagnosis: history, state of the art and perspective. Artif Intell Med 2001;23:89–109.

[4] Michie D, Spiegelhalter DJ, Taylor CC, editors. Machine learning neural and statistical classification. Ellis Horwood; 1994.

[5] Michalski RS, Bratko I, Kubat M, editors. Machine learning, data mining and knowledge discovery: methods and applications. John Wiley & Sons; 1998.

[6] Bellazzi R, Zupan B. Predictive data mining in clinical medicine: current issues and guidelines. Int J Med Inf 2008;77(2):81–97. doi: 10.1016/j.ijmedinf.2006.11.006. ISSN: 1386-5056.

[7] Cristianini N, Shawe-Taylor J. An introduction to support vector machines. Cambridge (UK): Cambridge University Press; 2000.

[8] Chapelle O, Vapnik V, Bousquet O, Mukherjee S. Choosing kernel parameters for support vector machines. AT&T Labs; 2000. Technical Report.

[9] Khan Y, Khan AA, Budiman FN, Beroual A, Malik NH, Al-Arainy AA. Partial discharge pattern analysis using support vector machine to estimate size and position of metallic particle adhering to spacer in GIS. Electr Pow Syst Res 2014;116:391–8. doi: 10.1016/j.epsr.2014.07.001. ISSN: 0378-7796.

[10] Altman NS. An introduction to kernel and nearest-neighbor nonparametric regression. Am Stat 1992;46(3):175–85.

CHAPTER SIX

Results, Discussions, and Research Challenges

6.1 RESULTS

In developed system, the methodology is adopted as proposed in previous chapters. The system consists of variety of steps starting from EEG signal acquisition to classification. This involves a wide subjective procedure, which gives us the idea that it is long-term process and output of one block depends on the further [1]. For example, if the signal is not preprocessed it may result in less classification accuracy. Hence, evaluation of each block is essential. Flowchart 6.1 describes the basic flowchart of the system.

Flowchart 6.1 shows the basic flow of the system used in research work. Firstly, raw EEG data are collected from the various sources (hospitals and some standard databases). Different EEG machines give different formats for reading the signals such as .edf, .eeg., .scn, and .csv which are some of the different formats found. In present research work, Recorders and Medicare Systems EEG machine was used to acquire the signals from patients. The EEG signal obtained from this machine was obtained in .eeg format. As the .eeg signal is unable to read in MATLAB, we have converted the same .eeg into .txt format. The text file contains the samples of EEG data. The .txt file is then imported into MATLAB for further preprocessing and features computations. The sample of EEG signal acquired using the RMS machine is shown below.

Fig. 6.1 shows the plot of EEG signal using EEGLAB and MATLAB. EEGLAB is a graphical user-interface tool, which can be used for preprocessing tool for EEG signals. The figure shows the plot of 21 channels' EEG signals plotted using the EEGLAB. EEGLAB is found to be an interactive MATLAB toolbox, which is used for processing the continuous and event-related EEG, MEG and other electrophysiological data using independent component analysis (ICA), time/frequency analysis and other methods including artifact rejection [2]. EEGLAB provides an interactive graphic user interface (GUI) allowing users to flexibly and interactively process their high-density EEG and other dynamic brain data using independent component analysis and/or time/frequency analysis (TFA), as well as standard averaging methods.

The placement of electrodes was done by using standard International 10/20 electrode placement system [2]. Accordingly, different electrodes play a significant role in diagnosis of Neurological disorders. In case of Alzheimer's disease, the Central (Cz), Parietal (Pz), Frontal (Fz) and Temporal (Tz) lobe plays a key role in diagnosis of

EEG-Based Diagnosis of Alzheimer Disease
http://dx.doi.org/10.1016/B978-0-12-815392-5.00006-X

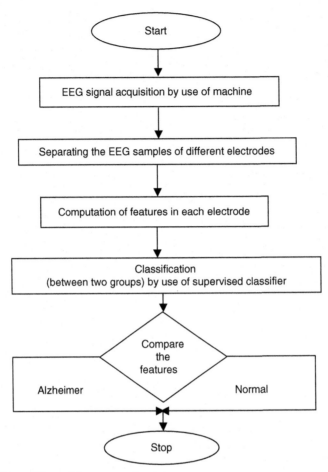

Flowchart 6.1 *General flow of the proposed system.*

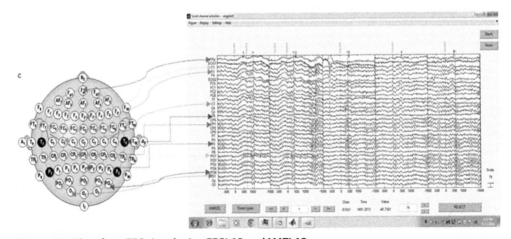

Figure 6.1 *Plot of raw EEG signal using EEGLAB and MATLAB.*

Alzheimer's disease. Hence, we have extracted the sample values of EEG signals in .txt format using RMS software system. Each EEG signal was then taken into the consideration for extracting the values of the each lobe of the signal. One hundred EEG signals were used for classification purpose. Similarly, different features were computed by considering the each signal as a row vector. At last, we obtained the feature matrix, suggesting that each column acts the features and rows contribute to the observations of each signal analyzed accordingly.

6.1.1 Spectral-based features

Several research findings have shown the changes in the EEG power spectra due to AD. It consists of increase in the delta and theta band powers, together with a decrease in alpha and beta band powers, thus suggesting a "Slowing of the EEG signal" [3–5]. We have measured the spectral power features present in each of the five conventional EEG frequency bands, namely: 0.1 − 4 Hz (delta), 4 − 8 Hz (theta), 8 − 12 Hz (alpha), 13 − 30 Hz (Beta) and 30 − 100 Hz (gamma). The above mentioned five spectral power features can be computed for per epoch for different 20 EEG electrodes and 8 bipolar channels. In our study, we have computed the same for above electrodes such as T3 (Temporal), F3 (Frontal), C3 (Central) and P4 (Parietal) electrodes, respectively. Thus Spectral Power based features play a significant role in diagnosing in Alzheimer Disease.

The raw EEG signal contains the unwanted frequency components and artifacts due to variety of interferences associated while recording of EEG signal. But, in order to analyze the EEG signal, we require the frequency of EEG signal from 0.5 Hz to 30 Hz. As such, these are the only frequency range of EEG signals in which ranges can be observed to distinguish between normal patients and patients with Alzheimer's disease. Hence, we have used the Band pass filter in order to filter out the signal frequency. Similarly, Independent Component Analysis (ICA) and Wavelet-based denoising techniques were also used to remove the certain artifacts associated in the signal [6]. The results of the above mentioned techniques are discussed in the previous chapters.

Flowchart 6.2 shows the computation of power features in each subband of EEG signals.

From Flowchart 6.2, it gives us the idea of calculating the Power based features in each subbands of each signals. Firstly, the filtered EEG signal is decomposed into the various bands using the wavelet decomposition tool. In this, we have used the "Daubechies" mother wavelet for decomposing the EEG signal into five different subbands. The reasons behind the use of Daubechies wavelet is justified in previous chapters. We have decomposed the EEG signal using "db2" Daubechies wavelet at level decomposition 5. Accordingly, we obtained the EEG signals into five bands with following frequencies Delta (0.5 − 4 Hz), Theta (4 − 8 Hz), Alpha (8-13 Hz) and Beta (13-30 Hz). Further, the Power in each subbands of EEG signal is computed by means of Power Density

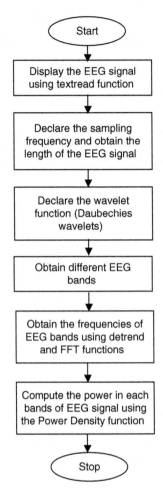

Flowchart 6.2 *Flowchart for computation of spectral-based features in EEG subbands.*

functions. Fig. 6.2 shows the outputs obtained after the computation of Spectral-based features.

Fig. 6.2 shows the plot of filtered EEG signal used for further analysis for extracting the different features.

Fig. 6.3 shows the classification of the EEG signals in various subbands calculated using the wavelet decomposition method. We have also used the Power Density function in order to compute the power in different bands of EEG signals in each electrode. It is observed that power in low frequency bands of patients infected with Alzheimer disease is increased whereas it is reduced in high frequency bands. Thus spectral-based feature helps out in diagnosis of AD (Figs. 6.4 and 6.5).

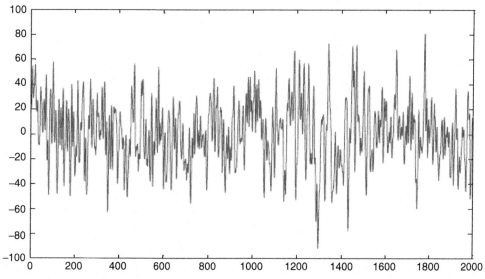

Figure 6.2 *Plot of EEG signal.*

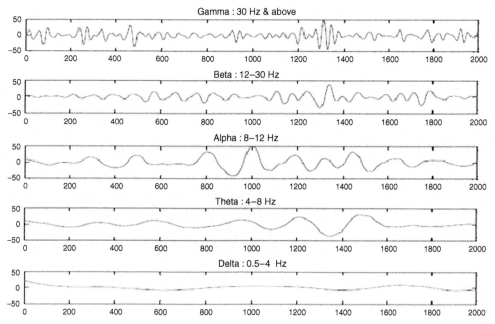

Figure 6.3 *Classification of EEG signals into different subbands.*

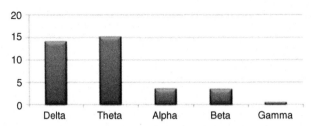

Figure 6.4 *Computations of relative power in EEG subbands in AD patients.*

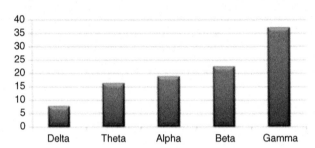

Figure 6.5 *Computations of relative power in EEG subbands in normal patients.*

Similarly, features values of each electrode of EEG signals were computed and then classified by use of suitable classifier. In case of wavelet-based features; mean and variance were computed of certain coefficient at level N.

Figs. 6.6 and 6.7 show the time frequency bumps observed in the EEG signal of the patients with Alzheimer's disease. In case of AD patients, the EEG signal shows the slowing effect due to the neuronal loss observed in the brain regions; but this phenomenon is not observed in the case of normal subjects. Dauwels et al. [5] have

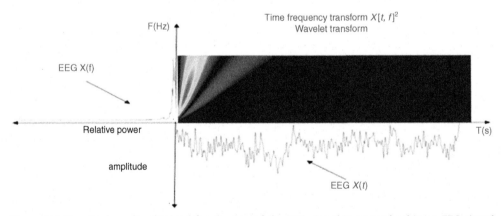

Figure 6.6 *Practical results obtained for slowing of the EEG signal in normal subjects: EEG signal, shown in time-domain x(t), frequency domain X(f), and time-frequency domain |X(t, f)|.*

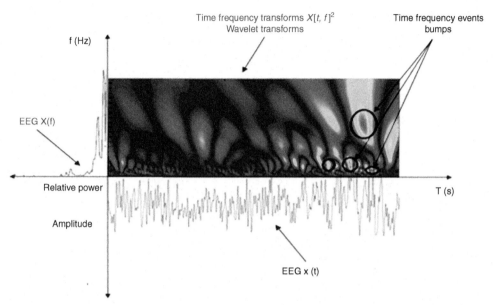

Figure 6.7 *Practical results obtained for slowing of the EEG signal in AD patients: EEG signal, shown in time-domain x(t), frequency domain X(f), and time-frequency domain |X (t,f)|.*

already justified this concept in his paper. But we have practically observed the bumps exhibited in case of our database. We also computed the relative power by means of time-frequency maps. Time-frequency maps of this EEG is quite sparse. Most energy is contained in specific regions of time frequency maps called as "bumps." It is observed that transient oscillations in the EEG signals of MCI and patients with Alzheimer's disease occur more often at low frequencies compared to the normal subjects. This is the signal of severe Alzheimer disease, in which the signal is slower. Thus cognitive deficits are tremendously affected in this stage. Fig. 6.7 also shows the time-frequency representation using wavelet transform; along with it, the relative power is also calculated which is decreased in delta and theta bands in the case of patients with Alzheimer's disease. These bumps are not observed in the case of normal patients since they do not exhibit slowing effect.

6.1.2 Wavelet-based features

In the present study, EEG signal is decomposed into various frequency bands by the use of wavelet decomposition technique. In the study, Daubechies wavelet is incorporated for calculating the various statistical features. We calculated the mean and variance in our study for calculating the various wavelet features. We have calculated the means and variance of decomposed coefficients of different lobes used in the study such as temporal and frontal lobes. In our study, we have used the "db2" mother wavelet at a level of

Table 6.1 Mean and variance of wavelet coefficients in temporal and frontal lobe in case of normal subjects

Sample no.	Temporal		Frontal	
	Mean	Var	Mean	Var
1	17.5466	648.872	202.445	18,409.2
2	3.27022	57,949.1	164.148	11,825.5
3	19.4064	52,375.7	−16.328	817.58
4	16.4304	9234.31	−7.544	449.301
5	170.527	9041.23	−17.874	1,203.15
6	43.7168	777.269	−4.9981	168.604
7	30.9504	6,058.95	16.978	122.833
8	43.7168	777.269	−376.99	60,616.7
9	−29.01	2,556.21	87.5291	4,068.15
10	−7.9323	640.901	−16.231	334.714
11	−29.01	2,556.21	31.0499	849.448
12	0.88173	308.146	−376.99	60,616.7
13	−153.35	12,849	5.32295	21.5091
14	−8.7402	574.171	−6.865	23.8506
15	1.05328	1,341.35	26.9282	530.71
16	39.119	634.427	5.32295	21.5091
17	23.3418	2,945.63	39.119	634.427
18	−33.261	791.999	−33.261	791.999
19	87.5291	4,068.15	87.5291	4,068.15
20	−16.231	334.714	−16.231	334.714

decomposition 5 in our work. Then these values are further extracted for classification purpose for distinguishing between two groups, that is, patients with Alzheimer's disease and normal subjects. The obtained values are listed in the following table.

Tables 6.1 and 6.2 show the computation of mean and variance of wavelet-based coefficients calculated for both normal as well as patients with Alzheimer's disease. From the observation of two tables, we can conclude that the values of mean and variance in the case of normal patients are high suggesting that the EEG of normal subjects is more complex as compared to Alzheimer's patients. This is due to the growth of neurons residing in the brain regions. Similarly, the value tends to be decreased in the case of patients with Alzheimer's disease due to the neuronal loss of brain cells or death of neurons in the cortical regions. This verifies that an EEG signal of EEG patients tends to be less complex as compared to that of the normal subjects.

6.1.3 Complexity-based features

Our study involved the use of different complexity-based features which were calculated in different lobes as mentioned above. The flowchart depicts the computation of complexity-based features used in the study.

Table 6.2 Mean and variance of wavelet coefficients in temporal and frontal lobe in case of patients with Alzheimer's disease

Sample no.	Temporal		Frontal	
	Mean	Var	Mean	Var
1	13.3374	79.864	39.1185	634.427
2	−33.98664	1,351.1421	23.341778	2,945.6273
3	−23.1839	236.334	−62.77812	7,396.1019
4	−4.191717	127.85384	−40.97355	492.79598
5	5.2668	360.72431	−622.3595	149,976.02
6	−4.197517	127.884	−32.2526	3,031.5972
7	−0.741	28,827.39	−17.6805	5,623.431
8	43.716755	777.26903	12.56433	2,158.0027
9	13.337461	79.864719	−19.05134	864.64996
10	42.383689	782.09568	−1.546154	172.60633
11	−33.98664	1,351.1421	−32.02303	593.99936
12	−7.53192	620.1975	−33.26098	791.99925
13	−33.98664	1,351.1421	10.97460	1,583.153
14	137.16467	7,971.852	−13.17483	1,080.4623
15	−7.53152	620.1975	70.12424	5,074.0601
16	16.978047	122.83285	28.922941	5,327.2333
17	−376.9918	60,616.737	−376.9918	60,616.737
18	87.529092	4,068.1513	87.529092	4,068.1513
19	−16.23056	334.71383	87.52902	4,068.1513
20	31.04996	849.44753	−16.23056	334.71383

The above Flowchart 6.3 shows the flowchart for computing the Complexity-based features. In our present work, we have computed four different complexity-based features such as Spectral Entropy, Spectral Roll-off, Spectral Centroid and Zero Crossing Rate for EEG signals. Firstly, the epoch of 2–3 seconds of EEG signals of about 150 samples was imported into the MATLAB. The signal was then filtered by means of Wavelet-based Denoising and Independent Component Analysis (ICA). The signals were then tested for Jarque Bera test to check whether they originate from normal distributions. MATLAB Statistical toolbox has an inbuilt function to check this condition. Hence, we have incorporated the use of same function to check this condition in our study. If the value of p is less than 0.001, then the signal is allowed to proceed for further analysis. But in our case, all signals were tested accordingly and satisfied the test. This test is mainly used for the analysis of Evoked Resting Potential (ERP- P300) EEG signals. In our case, all the signals used for training were tested, and they satisfied the Jarque Bera test.

After computing the about test, the signal computes the complexity features, and the values of each features are stored as features matrix. In this matrix, each column is one feature calculated for EEG sample, while each row is the observation for that particular

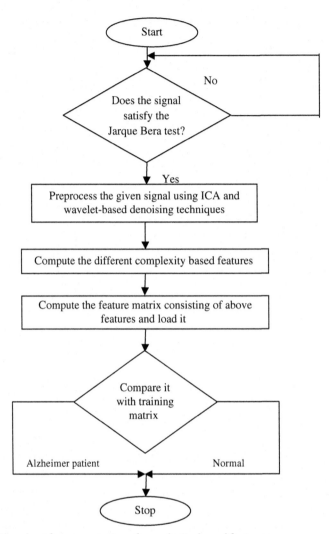

Flowchart 6.3 *Flowchart for computation of complexity-based features.*

feature. In this way, we obtain the feature matrix. After this, we check each EEG signal for predicting whether the given patient is suffering from AD. This is compared on the basis of classifier used. The classifier used contains two inputs; one is the feature matrix and other is the check matrix, which consists of the input signal to be checked. On the basis of comparison, the result of classifier is obtained for each sample. In our research work, we have used Support Vector Machine and K-Nearest Neighbor for classification. The following table shows the computation of various complexity-based features for different EEG signals in different lobes.

The results of the above features are shown in Table 6.3.

Table 6.3 Computation of complexity-based for central and frontal lobe for certain samples used in the study

EEG band/ feature	Central (C3)				Frontal (F4)			
	ZCR	SC	SR	SE	ZCR	SC	SR	SE
Sample 1	0.622	0.088	1.992	5.3981	0.0584	0.0863	1.9754	5.2619
Sample 2	**0.0012**	**0.0628**	**1.9321**	**5.95343**	**0.0018**	**0.0952**	**1.5423**	**2.8722**
Sample 3	**0.0916**	**0.1212**	**0.987**	**5.6044**	**0.0709**	**0.1102**	**1.9987**	**5.5409**
Sample 4	**0.01**	**0.0373**	**0.1322**	**3.644**	**0.0033**	**0.0437**	**0.9921**	**3.7955**
Sample 5	**0.0033**	**0.0489**	**0.0343**	**3.9189**	**0.033**	**0.0443**	**1.0932**	**3.9773**
Sample 6	0.09	0.1249	1.0323	5.7615	0.0967	0.1252	0.9812	5.0921
Sample 7	**0.0021**	**0.0032**	**0.0054**	**2.907**	**0.0034**	**0.087**	**0.878**	**3.0545**
Sample 8	0.09	0.1178	0.0324	4.897	0.0433	0.1245	0.2432	5.3429
Sample 9	**0.0167**	**0.1197**	**0.3432**	**3.5532**	**0.0167**	**0.0175**	**0.998**	**2.879**
Sample 10	0.0987	0.1526	1.98	6.006	0.0167	0.0956	1.9878	4.9978
Sample 11	0.0833	0.1417	1.9801	5.98	0.0433	0.1465	Na	5.9099
Sample 12	**0.067**	**0.1362**	**0.9802**	**4.4121**	**0.043**	**0.1321**	**0.889**	**4.1765**
Sample 13	0.0433	0.0981	1.9322	5.2557	0.033	0.0297	0.9121	3.4978
Sample 14	0.0567	0.1219	1.9212	5.681	0.0833	0.1185	1.0921	5.6421
Sample 15	0.0567	0.1274	0.1232	5.7301	0.0578	0.1167	0.9712	5.6117
Sample 16	0.682	0.1696	1.212	6.112	0.577	0.1392	1.992	5.895
Sample 17	0.0767	0.1045	1.889	5.3908	0.0367	0.0759	1.0923	4.9045
Sample 18	0.767	0.1232	1.921	5.543	0.0233	0.9832	1.7821	5.276
Sample 19	0.022	0.0516	1.821	4.1782	0.0567	0.1102	1.9922	5.5852
Sample 20	0.05	0.0927	1.212	5.3619	0.03	0.0935	0.9932	5.2706
Sample 21	0.09	0.1724	1.0932	6.2905	0.0833	0.1456	0.9912	5.9235
Sample 22	**0.0367**	**0.0756**	**0.9912**	**4.982**	**0.05**	**0.1379**	**0.9012**	**3.77812**
Sample 23	0.8434	0.778	1.232	6.345	0.232	0.343	0.9232	4.4556
Sample 24	**0.002**	**0.0054**	**0.767**	**3.454**	**0.021**	**0.0123**	**0.912**	**3.2232**
Sample 25	0.9912	0.8823	0.8832	4.556	0.676	0.0989	1.343	5.334

The samples in bold show the infected samples in Table 6.3. SC stands for Spectral Centroid, SE stands for Spectral Entropy, SR stands for Spectral Roll-off, and ZCR stands for Zero Crossing rate.

The samples in bold show the infected samples in Table 6.4. SC stands for Spectral Centroid, SE stands for Spectral Entropy, SR stands for Spectral Roll-off, and ZCR stands for Zero Crossing rate.

Fig. 6.8 shows the computation of complexity-based features for certain signals. The mean of each feature across all electrodes were calculated, and then they were plotted. Sample I had normal patients whereas Samples II and III had AD infected patients. From the chart plotted, we observe that higher values are obtained for each feature in the case of normal patients, which is highlighted by blue bar in the graph. Samples II and III were patients with Alzheimer's disease, which show the decreased values of different features.

Table 6.4 Computation of complexity-based for parietal and temporal lobe for certain samples used in the study

EEG band/ feature	Parietal (P4)				Temporal (T3)			
	ZCR	SC	SR	SE	ZCR	SC	SR	SE
Sample 1	0.0553	0.0744	1.9342	5.0385	0.042	0.074	0.9807	5.0385
Sample 2	**0.0343**	**0.0021**	**0.2311**	**2.0432**	**0.0143**	**0.055**	**0.9876**	**2.98**
Sample 3	**0.0229**	**0.0458**	**1.9928**	**4.1449**	**0.0756**	**0.1104**	**0.801**	**3.9801**
Sample 4	**0.0167**	**0.0363**	**0.9786**	**3.644**	**0.0032**	**0.032**	**0.8878**	**3.089**
Sample 5	**0.02**	**0.1921**	**1.0932**	**6.4532**	**0.143**	**0.564**	**2.323**	**5.5567**
Sample 6	0.0967	0.1946	1.565	6.4519	0.0367	0.0979	1.3433	5.3064
Sample 7	**0.0212**	**0.0343**	**1.0989**	**3.098**	**0.034**	**0.0023**	**0.879**	**3.9704**
Sample 8	0.09	0.1245	1.9878	5.4637	0.11	0.1595	1.9956	5.5765
Sample 9	**0.0367**	**0.0848**	**0.879**	**4.0177**	**0.0233**	**0.0824**	**1.3432**	**4.9456**
Sample 10	0.11	0.19	1.98	6.2752	0.0377	0.11	1.9801	5.5574
Sample 11	0.0833	0.1417	0.6743	5.869	0.0767	0.917	2.3214	5.0768
Sample 12	**0.0033**	**0.0554**	**1.231**	**3.8331**	**0.0033**	**0.0943**	**0.9565**	**4.4555**
Sample 13	0.0633	0.0987	1.0213	5.3167	0.03	0.0637	1.3221	4.6784
Sample 14	0.07	0.1159	0.3421	5.6396	0.0567	0.1219	1.9324	5.4312
Sample 15	0.0833	0.1035	0.5621	5.5579	0.0967	0.1313	0.5654	5.6529
Sample 16	0.0967	0.1561	1.3432	5.9423	0.13	0.1893	1.9922	6.1844
Sample 17	0.0033	0.0232	1.783	3.4554	0.099	0.1892	0.9321	3.0932
Sample 18	0.11	0.1256	1.9012	5.5643	0.0233	0.0243	1.9343	5.5654
Sample 19	0.0567	0.1217	1.9932	5.644	0.544	0.0834	1.922	5.095
Sample 20	0.0167	0.0877	1.9921	5.1347	0.0967	0.1491	1.0921	5.8923
Sample 21	0.11	0.1808	0.9912	6.2519	0.0833	0.1207	1.923	5.6688
Sample 22	**0.07**	**0.0732**	**1.0912**	**4.7801**	**0.05**	**0.0671**	**1.0921**	**4.7075**
Sample 23	0.3433	0.33212	0.4332	5.323	0.993	0.21	1.443	5.323
Sample 24	**0.0012**	**0.023**	**0.921**	**3.232**	**0.0443**	**0.354**	**0.844**	**3.545**
Sample 25	0.88	0.231	1.343	7.54	0.678	0.233	1.334	6.00

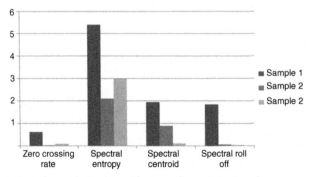

Figure 6.8 *Computation of complexity-based features for certain signals.*

This claims our hypothesis that EEG signal of patients with Alzheimer's disease tends to be less complex as compared to that of the normal subjects. This complexity is associated with the neurofibrillary tangles associated in the brain regions. It is also due to the neuronal loss of the brain cells that takes place in the cortical regions in the brain regions. Thus different complexity measures were studied and analyzed in the present study. Further, the different complexity features along with the spectral as well as wavelet-based features were classified for diagnosis using classifier. Along with this, we also calculated the diagnostic accuracy of the complexity-based features considering each single feature value taken into the consideration. We evaluated each of the features individually and obtained the following results by means of both the classifiers.

6.1.3.1 Support Vector Machine (SVM)
6.1.3.1.1 Zero Crossing Rate (ZCR)
Total patients considered for evaluation = 50 (consisting of 25 AD and 25 normal subjects).

Total trained data in database = 50 (consisting of both normal as well as patients with AD).

Correctly, classified AD patients = 21 and correctly classified normal patients = 20
Hence, total 45 patients are correctly classified.

The following are the results obtained after Classification, when only zero crossing rate features were taken as feature:
- Total number of Correctly identified AD individuals (TP) = 21
- Total number of Correctly identified normal individuals (TN) = 20
- Total number of misclassified AD individuals (FN) = 04
- Total number of misclassified normal individuals (FP) = 05
  Hence, accuracy = 21 + 20/23 + 22 + 3 + 2 = 42/50 = 84%.
  Thus diagnosis accuracy for patients with Alzheimer disease is given by,
  Accuracy = 21/25 = 82%

6.1.3.1.2 Spectral roll off (SR)
Total patients considered for evaluation = 50 (consisting of 25 AD and 25 normal subjects).

Total trained data in database = 50 (consisting of both normal as well as patients with AD).

Correctly, classified AD patients = 19 and correctly classified normal patients = 21
Hence, total 45 patients are correctly classified.

The following are the results obtained after classification, when only zero crossing rate features were taken as feature:
- Total number of correctly identified AD individuals (TP) = 19
- Total number of correctly identified normal individuals (TN) = 21.

- Total number of misclassified AD individuals (FN) = 06.
- Total number of misclassified normal individuals (FP) = 04

    Hence, accuracy = 19 + 21/23 + 22 + 3 + 2 = 40/50 = 90%.

    Thus diagnosis accuracy for patients with Alzheimer disease is given by, Accuracy = 19/25 = 76%

6.1.3.1.3 Spectral entropy (SE)

Total patients considered for evaluation = 50 (consisting of 25 AD and 25 normal subjects).

Total trained data in database = 50 (consisting of both normal as well as patients with AD).

Correctly, classified AD patients = 20 and correctly classified normal patients = 22 Hence, totally 45 patients are correctly classified.

The following are the results obtained after classification, when only zero crossing rate features were taken as feature:

- Total number of correctly identified AD individuals (TP) = 20
- Total number of correctly identified normal individuals (TN) = 22.
- Total number of misclassified AD individuals (FN) = 05.
- Total number of misclassified normal individuals (FP) = 03.

    Hence, accuracy = 20 + 22/50 = 42/50 = 84%.

    Thus diagnosis accuracy for patients with Alzheimer disease is given by accuracy = 20/25 = 80%

6.1.3.1.4 Spectral centroid (SC)

Total patients considered for evaluation = 50 (consisting of 25 AD and 25 normal subjects).

Total trained data in database = 50 (consisting of both normal as well as patients with AD).

Correctly, classified AD patients = 19 and correctly classified normal patients = 19 Hence, total 45 patients are correctly classified.

The following are the results obtained after classification when only zero crossing rate features were taken as feature:

- Total number of correctly identified AD individuals (TP) = 19
- Total number of correctly identified normal individuals (TN) = 19
- Total number of misclassified AD individuals (FN) = 06
- Total number of misclassified normal individuals (FP) = 06

    Hence, accuracy = 19 + 19/50 = 38/50 = 76%.

    Thus diagnosis accuracy for patients with AD is given by,

    Accuracy = 19/25 = 76%

    Thus we have obtained the following classification accuracy for Alzheimer diagnosis, which is shown in Fig. 6.9.

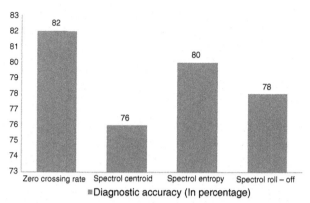

Figure 6.9 *Comparison of different complexity-based features used for predicting diagnostic accuracy using SVM classifier.*

6.1.3.2 K-Nearest neighbor classifier (K-NN)

6.1.3.2.1 Zero crossing rate (ZCR)

Total patients taken for evaluation = 50 (consisting of 25 AD and 25 normal subjects).

Total trained data in database = 50 (consisting of both normal as well as patients with Alzheimer's disease).

Correctly, classified AD patients = 18 and correctly classified normal patients = 19 Hence, total 45 patients are correctly classified.

The following are the results obtained after classification, when only zero crossing rate features were taken as feature:

- Total number of correctly identified AD individuals (TP) = 18.
- Total number of correctly identified normal individuals (TN) = 19.
- Total number of misclassified AD individuals (FN) = 07.
- Total number of misclassified normal individuals (FP) = 06

Hence, accuracy = 18 + 19/18 + 19 + 7 + 6 = 37/50 = 74%.

Thus diagnosis accuracy for patients with Alzheimer disease is given by, Accuracy = 18/25 = 72%

6.1.3.2.2 Spectral Roll-off (SR)

Total patients considered for evaluation = 50 (consisting of 25 AD and 25 normal subjects).

Total trained data in database = 50 (consisting of both normal as well as patients with Alzheimer's disease).

Correctly, classified AD patients = 18 and correctly classified normal patients = 20 Hence, a total of 45 patients are correctly classified.

The following are the results obtained after classification, when only zero crossing rate features were taken as feature:

- Total number of correctly identified AD individuals (TP) = 18.
- Total number of correctly identified normal individuals (TN) = 20.

- Total number of misclassified AD individuals (FN) = 07.
- Total number of misclassified normal individuals (FP) = 05.
  Hence, accuracy = 18 + 20/18 + 20 + 7 + 5 = 38/50 = 76%.
  Thus diagnosis accuracy for patients with Alzheimer disease is given by,
  Accuracy = 18/25 = 72%

6.1.3.2.3 Spectral entropy (SE)

Total patients taken for evaluation = 50 (consisting of 25 AD and 25 normal subjects).

Total trained data in database = 50 (consisting of both normal as well as patients with AD).

Correctly, classified AD patients = 20 correctly classified normal patients = 22
Hence, total 45 patients are correctly classified.

The following are the results obtained after classification when only zero crossing rate features were taken as feature:

- Total number of correctly identified AD individuals (TP) = 19
- Total number of correctly identified normal individuals (TN) = 20
- Total number of misclassified AD individuals (FN) = 06
- Total number of misclassified normal individuals (FP) = 05
  Hence, accuracy = 19 + 20/50 = 39/50 = 78%.
  Thus diagnosis accuracy for patients with Alzheimer disease is given by,
  Accuracy = 19/25 = 76%

6.1.3.2.4 Spectral centroid (SC)

Total patients considered for evaluation = 50 (consisting of 25 AD and 25 normal subjects).

Total trained data in database = 50 (consisting of both normal as well as patients with AD).

Correctly, classified AD patients = 18 and correctly classified normal patients = 17
Hence, a total of 45 patients are correctly classified.

The following are the results obtained after classification when only zero crossing rate features was taken as feature:

- Total number of correctly identified AD individuals (TP) = 18
- Total number of correctly identified normal individuals (TN) = 17
- Total number of misclassified AD individuals (FN) = 07
- Total number of misclassified normal individuals (FP) = 08
  Hence, accuracy = 18 + 17/50 = 35/50 = 70%.
  Thus diagnosis accuracy for patients with AD is given by
  Accuracy = 19/25 = 76%

From Figs. 6.9 and 6.10, we can observe that Zero crossing rate (ZCR) Spectral roll off (SR) increase the diagnostic accuracy in diagnosis of AD; whereas spectral centroid

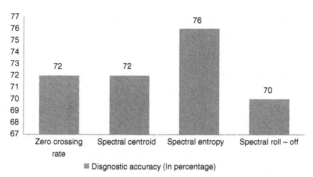

Figure 6.10 *Comparison of different complexity-based features used for predicting diagnostic accuracy using K-NN classifier.*

and spectral entropy show decrease in diagnostic accuracy. This can be due to signal of patients with Alzheimer's disease tends to be less complex and higher amount of spectral content is saturated in higher frequencies in the case of normal patients. The above-calculated features , that is, complexity features were proposed as the new set of features in the present study. The same features were tested, analyzed statistically and verified for the same results with the help of expert neurologists from different hospitals. The next section describes the classification output and diagnostic accuracy obtained in the research work.

6.1.4 Classification

In the present study, our main aim was to diagnose whether the given input signal of patient is normal or suffering from AD. This required a classifier to distinguish the two sample matrices as the input arguments along with one called as labels. This classification is done on the basis of different features used in previous study. We have incorporated the use of the supervised classification techniques. Supervised classification is one, which requires training and testing data for classification. Once a database is trained, testing of input data can be done efficiently. We have firstly trained the data of normal patients consisting of a check matrix as well as consisting of every features used. Once this training is finished, we can check for every signal for diagnosis between two groups. Flowchart 6.1 gives us the idea of classification involved in the research work. For every input signal imported, the analysis of the particular signal is done. It then starts comparing its calculated features with the trained database. The result whether the patient is normal or suffering from Alzheimer's disease is obtained on the basis of classifier used.

As already mentioned in previous chapters, we have involved the use of two supervised classifiers in our research work , that is, Support Vector Machine (SVM) and K-Nearest Neighbor (K-NN) classifier. Both the classifiers are efficiently used for the purpose of classification.

6.1.5 Calculation of diagnostic accuracy obtained in the research work

Based on the features calculated and classifier used, we have calculated the accuracy of classification based on following terminology.

$$Accuracy = \frac{(TP + TN)}{(TP + TN + FP + FN)} \qquad (6.1)$$

$$Sensitivity = \frac{TP}{(TP + FN)} \qquad (6.2)$$

$$Specificity = \frac{TN}{(TN + FP)} \qquad (6.3)$$

where *TP* stands for True Positive (AD individuals correctly classified), *TN* stands for True Negative (NC individuals correctly classified), *FP* stands for False Positives (NC individuals misclassified), *FN* stands for False Negative (AD individuals misclassified).

In our study, we have trained 50 EEG signals from Temporal, Frontal, Parietal and Central electrodes randomly. Out of which, the remaining 50 EEG signals were left out for testing comprising of both normal Alzheimer Affected persons. The following results were obtained after testing by the use of Support Vector machine (SVM classifier),

- Total number of correctly identified AD individuals (TP) = 24
- Total number of correctly identified normal individuals (TN) = 24
- Total number of misclassified AD individuals (FN) = 01
- Total number of misclassified normal individuals (FP) = 01
  Correspondingly, we have obtained the following results after calculating the values,
- Accuracy = (TP + TN)/(TP + TN + FP + FN) = 24 + 24 /(24 + 24 + 01 + 01) = 48/50 = 96%
- Sensitivity = TP/(TP + FN) = 24/(24 + 01) = 24/25 = 96%
- Specificity = TN/(FP + TN) = 24/ (22 + 03) = 24/25 = 96%
  Finally, we obtained the following results in our research work (Table 6.5).
  The following results were obtained after testing by use of K-Nearest neighbor classifier (K-NN classifier),
- Total number of correctly identified AD individuals (TP) = 23
- Total number of correctly identified normal individuals (TN) = 24
- Total number of misclassified AD individuals (FN) = 02

Table 6.5 Results indicating the accuracy obtained in the research work using SVM classifier

Accuracy	Sensitivity	Specificity	Computational time
96%	96%	96%	1.288837 s

Table 6.6 Results indicating the accuracy obtained in the research work using K-NN classifier

Accuracy	Sensitivity	Specificity	Computational time
94%	92%	96%	10.112 s

Table 6.7 Practical results obtained after classification

Classifier	Accuracy	Sensitivity	Specificity
SVM	96%	94%	96%
K-NN	94%	92%	96%

- Total number of misclassified normal individuals (FP) = 01
  Correspondingly, we have obtained the following results after calculating the values
- Accuracy = (TP + TN)/(TP + TN + FP + FN) = 23 + 24 /(23 + 24 + 01 + 01) = 47/50 = 94%
- Sensitivity = TP/(TP + FN) = 23/(24 + 01) = 23/25 = 92%
- Specificity = TN/(FP + TN) = 24/(22 + 03) = 24/ 25 = 96%
  Finally, we obtained the following results in our research work (Table 6.6).
  Let us compare the final results obtained after classification done by both supervised learning algorithms (Table 6.7).
  Fig. 6.11 and result as shown in Table 6.8 show that SVM classifier gives out good classification accuracy as compared to that of the K-NN classifier. It is due to SVM classifier, which classifies each input data efficiently in the case of high dimensional data. It is also verified that SVM classifier takes less computational time for the classification of input data (Fig. 6.12).
  Fig. 6.12 shows the diagnostic accuracy obtained by using various methods for the diagnosis of Alzheimer's disease by means of EEG signals. It is observed that the complexity-based features play a significant role in the diagnosis of Alzheimer's disease

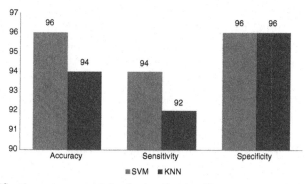

Figure 6.11 *Classification accuracy obtained using the classifiers.*

Table 6.8 Computational time required for execution of each parameter

Sr. No.	Illustration	Time (in seconds)
1	Time required to execute training data	36.121994
2	Time required to execute testing of sample data using SVM classifier	1.288837
3	Time required to execute testing of sample data using K-NN classifier	10.112

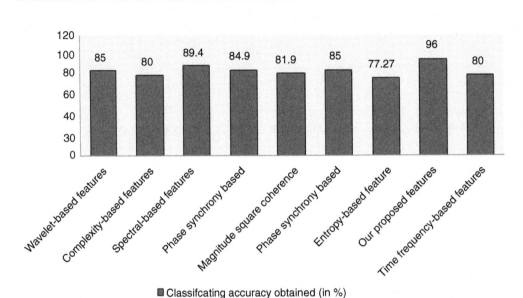

■ Classifcating accuracy obtained (in %)

Figure 6.12 *Comparison of different features used for increase in diagnostic accuracy.*

giving classification accuracy of 96%; which is much higher as compared to previous studies.

6.2 CONCLUSIONS

This chapter presents new signal processing and machine learning methods for the early diagnose of AD using EEG signals. The main steps used in EEG data analysis have been explored, from the preprocessing to the classification. The main conclusions drawn from this study are presented below.

Preprocessing is an important step, which facilitates the posterior analysis of the data. Two different cleaning methods namely Independent Component Analysis (ICA) Wavelet-based Denoising have been presented. These cleaning methods have been evaluated on simulated data and on real data. Results presented using simulated data with different levels show that for signals presenting high level of noise and artifacts, the cleaning method improves the quality of the data when compared to data without artifacts. However, for signals with a low level of noise or artifacts, the cleaning method is not as effective. Results obtained after using the cleaning methods on real data presents the same characteristics, obtaining a small classification improvement when compared with data without cleaning. However, as pointed in previous chapters, in order to apply this method in real data an important step has to be added. This pre-cleaning step would be used to decide if the data are clean enough, avoiding unnecessary cleaning.

Different biomarkers indicative of changes that AD causes on EEG data have been identified. Afterward, these measures have been used individually to distinguish between healthy subjects and AD patients in different stages (MCI and Mild AD). Different features have been explored in the present study such as Spectral-based features, Wavelet-based features and Complexity-based features. We have concluded that power in low frequency bands of EEG signals such as Delta (0-4.5 Hz) and Theta (4.5-8 Hz) increases, while power in high frequency bands such as Alpha (8-12 Hz) and Beta (13-30 Hz) decreases in the case of patients with Alzheimer's disease due to the neuronal loss of cells and neurofibrillary tangles associated with the brain cells. Similarly, we have used the wavelet-based features to distinguish between two groups. In this study, we calculated the statistical values of the signal after denoising it using wavelet transform , that is, mean and standard deviation. It shows that EEG signal of patients with Alzheimer's disease is less complex as compared to that of the normal subjects. This is associated due the neuronal loss and death of neurons in the brain cells [7].

In present study, our main aim was to investigate and observe the effects of different complexity-based features on EEG signals of both Alzheimer Disease normal patients. In literature, as already seen in Chapter 2, we can observe that Spectral-Based features such as EEG Relative Power, Magnitude Square Coherence, Phase Synchrony and EEG amplitude Modulation Energy are widely used, which play a significant role in AD diagnosis giving accuracy of about more than 80%. Our hypothesis in this study was to prove that EEG of AD patients tend to be less complex as compared to the normal subjects due to neuronal loss of cells in brain regions by the use of this complexity features such as Spectral Centroid, Spectral Roll-off, Spectral Entropy and Zero Crossing rate.

On the basis of the above results and features used, we have evaluated the different EEG-based complexity measures to observe and study them if they carry any diagnostic useful information for the diagnosis of Alzheimer disease. In medical concept, it is signified that AD affects the neuronal activity of the patients. In this study, we have claimed the hypothesis that EEG signals of AD patients have less signal complexity as that compared to the CN subjects. The above used features show decreased complexity values for AD patients, which practically confirms our hypothesis. The difference in the complexity-based feature values among the cohort is small, but indicates its significance on the electrodes of EEG. The AD group features consist of lower values, suggesting that AD subjects tend to be less complex. The features used to carry relevant information in the central, parietal, temporal and frontal lobes. This reduced complexity occurs due to the appearance of the neurofibrillary plaques and tangles as already discussed.

Spectral Entropy, Spectral Roll – off, Spectral Centroid and Zero Crossing Rate values were also lower for patients with Alzheimer disease in the frontal and temporal lobes. In this study, we obtained 96% classification accuracy using SVM classifier and 94% accuracy by the use of K-NN classifier. It is observed that there exists a higher amount of spectral content in higher frequencies for CN group. This is predicted as the high level of complexity in CN subjects. In this manner, our hypothesis is verified.

It is to highlight that genetic modification of stem cells in AD patients also makes the slowing of EEG signals more regular. It is also observed that neural connectivity gets increased in the brain cells and Aβ (Beta Amyloid) protein gets degenerated. The stem cell increases the level of neprilysin. It is an enzyme that breaks the level of Aβ protein and lowers the brain activity of AD patients [8]. Hence, an alpha and beta activity also reduces, and complexity gets reduced as the signal becomes slow.

6.3 CONTRIBUTION

Real time EEG signals were collected with consultation of doctors and expert neurologists from different hospitals in Pune, India such as Jagtap Clinic and Research Centre, and Smt. Kashibai Navale Medical College and General Hospital. The EEG signals were studied in detail with the help of neurologists. Along with it, the EEG setup was also studied, and an EEG machine was also studied during recording of real time data. Different EEG signals were analyzed firstly with consultation of neurologists to understand the difference between normal and patients with Alzheimer's disease. Different types of EEG signals such as ERP signals were also studied, which are mainly used in Cognitive Neuroscience for the study of neurodegenerative diseases such as Alzheimer's disease and Epilepsy. But such a kind of ERP signals were not used in our study due to the shortage of ERP signals because it requires auditory system for recording of ERP signals. In this study, we have claimed the hypothesis that EEG signals of Alzheimer's disease are less complex as compared to that of normal patients. In this study, our aim was to investigate and observe the effects of different complexity-based features on EEG signals of both Alzheimer Disease and normal patients. In this study, we also evaluated each complexity-based feature individually and observed that Zero crossing rate (82%) and Spectral Roll-off (76%) give us more diagnostic accuracy as compared to that of the other two features. But, when we combine all these features; we obtain a good and satisfactory diagnostic accuracy (96%). We have observed the use of same features to verify our hypothesis and verified it accordingly.

6.4 LIMITATIONS OF THE STUDY

In this study, we have investigated the use of complexity-based features for the diagnosis of Alzheimer's disease using EEG signals. But, same features can be verified for the diagnosis of Alzheimer's disease for ERP signals by performing some cognitive actions. The acquired real time EEG signals contain different artifacts due to eye blinking, muscle activity and other machine artifacts such as Power line interferences and so on. These kinds of artifacts can also be eliminated by the use of different artifact removal algorithms. This limitation should be overcome in order to increase the classification accuracy or diagnostic accuracy. We have verified our results for a total of 100 samples; such a dataset must be increased in order to obtain more precise diagnostic accuracy.

6.5 FUTURE SCOPE

Our future work in this study includes the automated diagnosis and classification of EEG data using various classifiers including Neural Networks, Support Vector Machines etc., to increase the diagnostic accuracy for distinguishing between AD and CN group. In this study, we investigated the use of complexity-based features for automated diagnosis of AD. It is to highlight that when we combine these features with one another, they provide more diagnostic information. Our future work also extends to improve the quality of EEG signal by the use of automated algorithms in order to increase the classification accuracy. The work can be patented. The future plans include the more reliable system for greater up to 100% accuracy of the signal for detection of the Alzheimer disease. The system can also be made portable in future. The EEG signals show some of the radiations effects, which are harmful for the patients' life. The minimization of the radiation effects is also a future scope of the research.

Future work in this study involves making analysis of each frequency bands in depth to observe whether they carry any other significant information for better diagnosis by means of various signal processing algorithms and features extraction techniques. Our future work will also include implementing present algorithms on hardware devices such as DSP processors (Application Specific Integrated Circuits) and FPGA (Field Programmable Gate Array) devices to make a standalone device for diagnosis, which might be useful for doctors for correctly diagnosing the patients in early stage [8]. This work explores new tool for Alzheimer disease diagnosis and further research using some more of these features can report remarkable achievements in this field.

REFERENCES

[1] Saima F, Muhammad Abuzar F, Huma T. An ensemble-of-classifiers based approach for early diagnosis of Alzheimer's Disease: classification using structural features of brain images. Comp Math Meth Med 2014;2014:11. doi: 10.1155/2014/862307. Article ID 862307.
[2] Nilesh Kulkarni N, Bairagi VK. Diagnosis of Alzheimer disease using EEG signals. Int J Eng Res Technol (IJERT) 2014;3(4):1835–8. Apr.
[3] Jeong J. EEG dynamics in patients with Alzheimer's disease. Clin Neurophysiol 2004;15(7):1490–505.
[4] Dauwels J, Srinivasan K, Ramasubba Reddy M, Musha T, Vialatte F-B, Latchoumane C, Jeong J, Cichocki A. Slowing and loss of complexity in Alzheimer's EEG: two sides of the same coin? Int J Alzheimers Dis 2011;2011:539621.
[5] Dauwels J, Vialatte F, Cichocki A. Diagnosis of Alzheimer's disease from EEG signals: where are we standing? Curr Alzheimer Res 2010;7(6):487–505. Sep.

[6] Kulkarni NN, Bairagi VK. Extracting salient features for EEG based diagnosis of Alzheimer disease using support vector machine classifier. IETE J Res 2017;63(1):11–22. Jan.

[7] Tyler S, Robi P. Analysis of complexity-based EEG features for diagnosis of Alzheimer disease. Proc Intl Conf IEEE-EMBC; 2011Boston, USA. p. 2033–6.

[8] Blurton-Jones M, Spencer B, Michael S, Castello NA, Agazaryan AA. Neural stem cells genetically-modified to express neprilysin reduce pathology in Alzheimer transgenic models. Stem Cell Res. Ther. 2014;5(46)doi: 10.1186/scrt440. Apr.

INDEX

Printed in the United States
By Bookmasters